Advanced Instrumentation and Computer I/O Design

Defined Accuracy Decision and Control with Process Applications

Second Edition

Patrick H. Garrett

IEEE Press

A John Wiley & Sons, Inc., Publication

P9-DUB-657

Copyright © 2013 by the Institute of Electrical and Electronics Engineers, Inc.

Published by John Wiley & Sons, Inc., Hoboken, New Jersey. All rights reserved.
Published simultaneously in Canada.

No part of this publication may be reproduced, stored in a retrieval system or transmitted in any form or by any means, electronic, mechanical, photocopying, recording, scanning or otherwise, except as permitted under Section 107 or 108 of the 1976 United States Copyright Act, without either the prior written permission of the Publisher, or authorization through payment of the appropriate per-copy fee to the Copyright Clearance Center, Inc., 222 Rosewood Drive, Danvers, MA 01923, (978) 750-8400, fax (978) 750-4470, or on the web at www.copyright.com. Requests to the Publisher for permission should be addressed to the Permissions Department, John Wiley & Sons, Inc., 111 River Street, Hoboken, NJ 07030, (201) 748-6011, fax (201) 748-6008, or online at http://www.wiley.com/go/permission.

Limit of Liability/Disclaimer of Warranty: While the publisher and author have used their best efforts in preparing this book, they make no representation or warranties with respect to the accuracy or completeness of the contents of this book and specifically disclaim any implied warranties of merchantability or fitness for a particular purpose. No warranty may be created or extended by sales representatives or written sales materials. The advice and strategies contained herein may not be suitable for your situation. You should consult with a professional where appropriate. Neither the publisher nor author shall be liable for any loss of profit or any other commercial damages, including but not limited to special, incidental, consequential, or other damages.

For general information on our other products and services please contact our Customer Care Department within the United States at (800) 762-2974, outside the United States at (317) 572-3993 or fax (317) 572-4002.

Wiley also publishes its books in a variety of electronic formats. Some content that appears in print, however, may not be available in electronic formats. For more information about Wiley products, visit our web site at www.wiley.com.

Library of Congress Cataloging-in-Publication Data:

Garrett, Patrick H.
  Advanced instrumentation and computer I/O design : defined accuracy decision and control with process applications / Patrick H. Garrett. — Second edition.
    pages cm
    ISBN 978-1-118-31708-2 (pbk.)
1. Computer interfaces. 2. Computer input-output equipment—Design—Data processing. 3. Computer-aided engineering. I. Title.
  TK7887.5.G368 2013
  621.39'81—dc23   2012030698

Printed in the United States of America.

10 9 8 7 6 5 4 3 2 1

Library
University of Texas
at San Antonio

Advanced Instrumentation and Computer I/O Design

WITHDRAWN
UTSA LIBRARIES

IEEE Press
445 Hoes Lane
Piscataway, NJ 08854

IEEE Press Editorial Board
John B. Anderson, *Editor in Chief*

Ramesh Abhari	Bernhard M. Haemmerli	Saeid Nahavandi
George W. Arnold	David Jacobson	Tariq Samad
Flavio Canavero	Mary Lanzerotti	George Zobrist
Dmitry Goldgof	Om P. Malik	

Kenneth Moore, *Director of IEEE Book and Information Services (BIS)*

Contents

Preface

Widespread need continues across aerospace, biomedical, commercial, and federal domains for the systematic design of instrumented processes aided by advanced decision-and-control methodologies. Realizations have evolved beneficially incorporating defined accuracy data with process automation designs that enable effective attainment of system goals with nominal variability. In this book, real-world applications are accordingly developed, illustrated by a dozen case studies performed for technology enterprises, including the Air Force Materials and Manufacturing Directorate, General Electric Aviation, General Motors Technical Center, Goodyear Tire & Rubber, U.S. Environmental Protection Agency, and Wheeling-Pittsburgh Steel.

The initial sixty percent of this book consecutively develops from input sensor signal conditioning to output sampled-data linear signal reconstruction designs, at data accuracies of interest, limited primarily by the residual errors of included electronic devices. Real-time computer I/O systems are traditionally circuit-design based. Defined accuracy I/O designs alternatively employ device and system models developed in Chapters 1 through 6, as concentrated by the featured instrumentation analysis suite workbook of Chapter 7, that includes a modifiable user interface to exercise chosen device data and system parameters for the evaluation of end-to-end system signal accuracy. This may be downloaded at http://booksupport.wiley.com.

The remaining forty percent of the book evolves process design methods culminating in a hierarchical subprocess control architecture. This includes an upper-level feedforward planner containing models of product features that outputs references to intermediate-level in-situ subprocesses, enabling more definitive feedback control, separate from physical apparatus regulation. Processing effectiveness is achieved by means of focused subprocess decoupling, sensor fusion, accountable data attribution, sensed process migration control planner remodeling, and computational-intelligence reasoning when quantitative models are incomplete. End-of-chapter problems with separate solutions are included as exercises.

PATRICK H. GARRETT
Automation Center
University Of Cincinnati

CHAPTER 1

Thermal, Mechanical, Quantum, and Analytical Sensors

1-0 INTRODUCTION

Automated laboratory systems, manufacturing process controls, analytical instrumentation, and aerospace systems all would have diminished capabilities without the availability of contemporary computer-integrated data systems with multisensor information structures. This text accordingly develops supporting error models that enable a unified performance evaluation for the design and analysis of linear and digital instrumentation systems with the goal of compatibility of integration with enterprise quality requirements. These methods then serve as a quantitative framework supporting the design of high-performance automation systems.

This chapter specifically describes the front-end electrical sensor devices for a broad range of applications from industrial processes to scientific measurements. Examples include environmental sensors for temperature, pressure, level, and flow; optical sensors for measurements beyond apparatus boundaries, including spectrometers for chemical analytes; and material and biomedical assays sensed by microwave microscopy. It is notable that owing to advancements in higher attribution sensors they are increasingly being substituted for process models in many applications.

1-1 INSTRUMENTATION ERROR INTERPRETATION

Measured and modeled electronic device, circuit, and system error parameters are defined in this text for combination into a quantitative end-to-end instrumentation

Advanced Instrumentation and Computer I/O Design, Second Edition. By Patrick H. Garrett
Copyright © 2013 the Institute of Electrical and Electronics Engineers, Inc.

1

performance representation for computer-centered measurement and control systems. It is therefore axiomatic that the integration and optimization of these systems may be achieved by design realizations that provide total error minimization. Total error is graphically described in Figure 1-1, and analytically expressed by Equation (1-1), as a composite of mean error contributions (barred quantities) plus the root-sum-square (RSS) of systematic and random uncertainties. Total error thus constitutes the deviation of a sensor-based measurement from its absolute true value, which is traceable to a standard value harbored by the National Institute of Standards and Technology (NIST). This error is traditionally expressed as 0–100% of full scale (%FS), where the RSS component represents a one-standard-deviation confidence interval, and accuracy is defined as the complement of error ($100\% - \varepsilon_{\text{total}}$). Figure 1-2 illustrates generic sensor elements and the definitions describe relevant terms:

Accuracy: The closeness with which a measurement approaches the true value of a measurand, usually expressed as a percent of full scale

Error: The deviation of a measurement from the true value of a measurand, usually expressed as a percent of full scale

Tolerance: Allowable deviation about a reference of interest

Precision: An expression of a measurement over some span described by the number of significant figures available

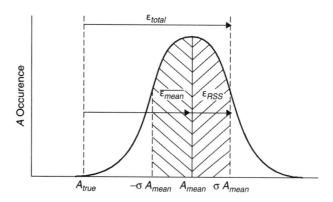

Figure 1-1. Instrumentation error interpretation.

$$\varepsilon_{\text{total}} = \sum \overline{\varepsilon_{\text{mean}}}\,\%\text{FS} + \left[\sum \varepsilon_{\text{systematic}}^2 + \sum \varepsilon_{\text{random}}^2\right]^{1/2}\%\text{FS}1\sigma \qquad (1\text{-}1)$$

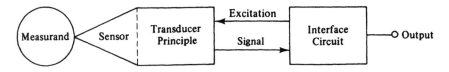

Figure 1-2. Generic sensor elements.

Resolution: An expression of the smallest quantity to which a quantity can be represented

Span: An expression of the extent of a measurement between any two limits

Range: An expression of the total extent of measurement values

Linearity: Variation in the error of a measurement with respect to a specified span of the measurand

Repeatability: Variation in the performance of the same measurement

Stability: Variation in a measurement value with respect to a specified time interval

Technology has advanced significantly as a consequence of sensor development. Sensor nonlinearity is a common source of error that can be minimized by means of multipoint calibration. Practical implementation often requires the synthesis of a linearized sensor that achieves the best asymptotic approximation to the true value over a measurement range of interest.

The cubic function of Equation (1-2) is an effective linearizing equation demonstrated over the full 700°C range of a commonly applied Type-J thermocouple, which is tabulated in Table 1-1. Solution of the A and B coefficients at judiciously spaced temperature values defines the linearizing equation with a 0°C intercept. Evaluation at linearized 100°C intervals throughout the thermocouple range reveals temperature values nominally within 1°C of their true temperatures, which correspond to typical errors of 0.25%FS. It is also useful to express the average of discrete errors over the sensor range, obtaining a mean error value of $\overline{0.11}$%FS for the Type-J thermocouple. This example illustrates a design goal proffered throughout this text of not exceeding one-tenth percent error for any contributing system component. Extended polynomials permit further reduction in linearized sensor error while incurring increased computational burden, where a fifth-order equation can beneficially provide linearization to 0.1°C, corresponding to $\overline{0.01}$%FS mean error.

Table 1-1. Sensor cubic linearization

Y °C	X mV	y °C	$\varepsilon_{\%FS} = |(Y-y)100\%/700°C|$
0	0	0	0
100	5.269	98	0.27
200	10.779	200	0
300	16.327	302	0.25
400	21.848	401	0.23
500	27.393	500	0
600	33.102	599	0.17
700	39.132	700	0

Y = true temperature
X = Type-J thermocouple signal
y = linearized temperature
$\overline{0.11\%FS}$ mean error outcome for 0°C intercept with 700°C full scale range

$$y = AX + BX^3 + \text{intercept} \qquad (1\text{-}2)$$

Coefficient for 10.779 mV at 200°C:

$$200°C = A(10.779 \text{ mV}) + B(10.779 \text{ mV}^3) + 0°C$$

$$A = 18.5546 \frac{°C}{mV} - B(116.186 \text{ mV}^2)$$

Coefficient for 27.393 mV at 500°C:

$$500°C = 508.2662 °C - B(3182.68 \text{ mV}^3) + B(20{,}555.0 \text{ mV}^3)$$

$$A = 18.6099$$

$$B = -0.000475 \frac{°C}{mV^3}$$

1-2 TEMPERATURE SENSORS

Thermocouples are widely used temperature sensors because of their ruggedness and broad temperature range. Two dissimilar metals are used in the Seebeck-effect temperature-to-emf junction with transfer relationships described by Figure 1-3. Proper operation requires the use of a thermocouple reference junction in se-

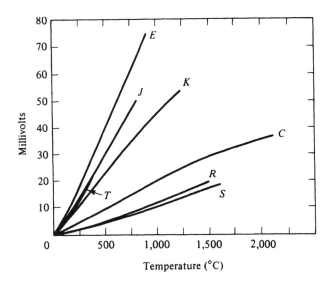

Figure 1-3. Temperature–millivolt graph for thermocouples. (Courtesy Omega Engineering, Inc., an Omega Group Company.)

ries with the measurement junction to polarize the direction of current flow and maximize the measurement emf. Omission of the reference junction introduces an uncertainty evident as a lack of measurement repeatability equal to the ambient temperature.

An electronic reference junction that does not require an isolated supply can be realized with an Analog Devices AD590 temperature sensor as shown in Figure 4-5. This reference junction usually is attached to an input terminal barrier strip in order to track the thermocouple-to-copper circuit connection thermally. The error signal is referenced to the Seebeck coefficients in mV/°C of Table 1-2,

Table 1-2. Thermocouple comparison data

Type	Elements, +/–	mV/°C	Range (°C)	Application
E	Chromel/constantan	0.063	0 to 800	High output
J	Iron/constantan	0.054	0 to 700	Reducing atmospheres
K	Chromel/alumel	0.040	0 to 1,200	Oxidizing atmospheres
R&S	Pt–Rh/platinum	0.010	0 to 1,400	Corrosive atmospheres
T	Copper/constantan	0.040	–250 to 350	Moist atmospheres
C	Tungsten/rhenium	0.012	0 to 2,000	High temperature

and provided as a compensation signal for ambient temperature variation. The single calibration trim at ambient temperature provides temperature tracking with a few tenths of a °C.

Resistance-thermometer devices (RTDs) provide greater resolution and repeatability than thermocouples, the latter typically being limited to approximately 1°C. RTDs operate on the principle of resistance change as a function of temperature, and are represented by a number of devices. The platinum resistance thermometer is frequently utilized in industrial applications because it offers accuracy with mechanical and electrical stability. Thermistors are fabricated from a sintered mixture of metal alloys, forming ceramics exhibiting a significant negative temperature coefficient. Metal film resistors have an extended and more linear range than thermistors, but thermistors exhibit approximately ten times the sensitivity. RTDs require excitation, usually provided as a constant-current source, in order to convert their resistance change with temperature into a voltage change. Figure 1-4 presents the temperature–resistance characteristics of common RTD sensors.

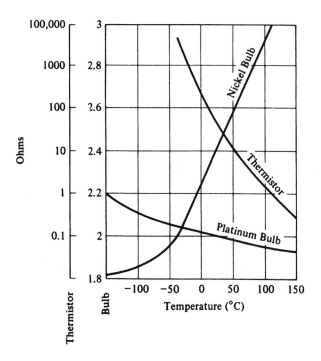

Figure 1-4. RTD devices.

Optical pyrometers are utilized for temperature measurement when sensor physical contact with a process is not feasible, but a view is available. Measurements are limited to energy emissions within the spectral response capability of the specific sensor used. A radiometric match of emissions between a calibrated reference source and the source of interest provides a current analog corresponding to temperature. Automatic pyrometers employ a servo loop to achieve this balance, as shown in Figure 1-5. Operation to 5000°C is available.

1-3 MECHANICAL SENSORS

Fluid pressure is defined as the force per unit area exerted by a gas or a liquid on the boundaries of a containment vessel. Pressure is a measure of the energy content of hydraulic and pneumatic (liquid and gas) fluids. Hydrostatic pressure refers to the internal pressure at any point within a liquid directly proportional to the liquid height above that point, independent of vessel shape. The static pressure of a gas refers to its potential for doing work, which does not vary uniformly with height as a consequence of its compressibility. Equation (1-3) expresses the basic relationship between pressure, volume, and temperature as the general gas law. Pressure typically is expressed in terms of pounds per square inch (psi) or inches of water (in H_2O) or mercury (in Hg). Absolute pressure measurements are refer-

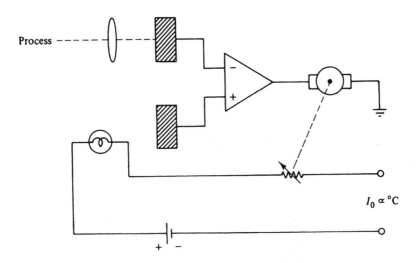

Figure 1-5. Automatic pyrometer.

enced to a vacuum, whereas gauge pressure measurements are referenced to the atmosphere.

A pressure sensor responds to pressure and provides a proportional analog signal by means of a pressure–force summing device. This usually is implemented with a mechanical diaphragm and linkage to an electrical element such as a potentiometer, strain gauge, or piezoresistor. Quantities of interest associated with pressure–force summing sensors include their mass, spring constant, and natural frequency. Potentiometric elements are low in cost and have high output, but their sensitivity to vibration and mechanical nonlinearities combine to limit their utility. Unbonded strain gauges offer improvements in accuracy and stability, with errors to 0.5% of full scale, but their low output signal requires a preamplifier. Present developments in pressure transducers involve integral techniques to compensate for the various error sources, including crystal diaphragms for freedom from measurement hysteresis. Figure 1-6 illustrates a microsensor-circuit pressure transducer for enhanced reliability, with an internal vacuum reference, chip heating to minimize temperature errors, and a piezoresistor bridge transducer circuit with on-chip signal conditioning.

$$\frac{\text{Absolute pressure} \times \text{Gas volume}}{\text{Absolute temperature}} = \text{Constant} \qquad (1\text{-}3)$$

Liquid levels are frequently required process measurements in tanks, pipes, and other vessels. Sensing methods of various complexity are employed, includ-

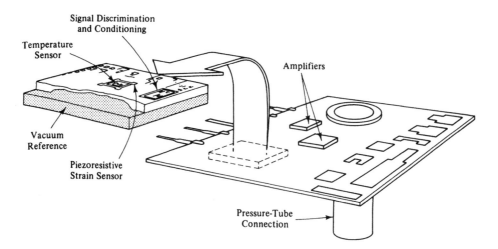

Figure 1-6. Integrated pressure microsensor.

ing float devices, differential pressure, ultrasonics, and bubblers. Float devices offer simplicity and various means of translating motion into a level reading. A differential-pressure transducer can also measure the height of a liquid when its specific weight W is known, and a ΔP cell is connected between the vessel surface and bottom. Height is provided by the ratio of $\Delta P/W$.

Accurate sensing of position, shaft angle, and linear displacement is possible with the linear variable-displacement transformer (LVDT). With this device, an ac excitation introduced through a variable-reluctance circuit is induced in an output circuit through a movable core that determines the amount of displacement. LVDT advantages include overload capability and temperature insensitivity. Sensitivity increases with excitation frequency, but a minimum ratio of 10:1 between excitation and signal frequencies is considered a practical limit. LVDT variants include the induction potentiometer, synchros, resolvers, and the microsyn. Figure 1-7 describes a basic LVDT circuit with both ac and dc outputs.

Fluid-flow measurement generally is implemented either by differential-pressure or mechanical-contact sensing. Flow rate F is the time rate of fluid motion, with dimensions typically in feet per second. Volumetric flow Q is the fluid volume per unit time, such as gallons per minute. Mass flow rate M for a gas is defined, for example, in terms of pounds per second. Differential-pressure-flow-sensing elements also are known as variable-head meters because the pressure difference between the two measurements ΔP is equal to the head. This is equiva-

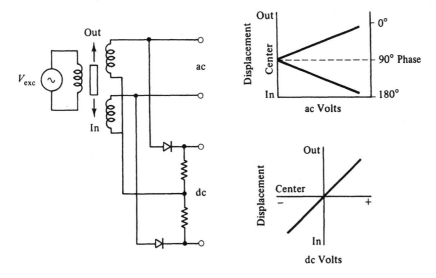

Figure 1-7. Basic LVDT.

lent to the height of the column of a differential manometer. Flow rate is therefore obtained with the 32 ft/sec^2 gravitational constant g and differential pressure by Equation (1-4). Liquid flow in open channels is obtained by head-producing devices such as flumes and weirs. Volumetric flow is obtained with the flow cross-sectional area and the height of the flow over a weir, as shown by Figure 1-8 and Equation (1-5).

$$\text{Flow rate } F = \sqrt{2g\Delta P} \text{ feet/second} \qquad (1\text{-}4)$$

$$\text{Volumetric flow } Q = \sqrt{2gL^2H^3} \text{ cubic feet/second} \qquad (1\text{-}5)$$

$$\text{Mass flow } M = \sqrt{R\frac{\Delta P_o}{\Delta P_x}} \cdot \sqrt{\frac{P\Delta P}{T}} \text{ pounds/second} \qquad (1\text{-}6)$$

where

R = universal gas constant

ΔP_o = true differential pressure $P_o - P_\infty$

ΔP_x = calibration differential pressure

Acceleration measurements are principally of interest for shock and vibration sensing. Potentiometric dashpots and capacitive transducers have largely been supplanted by piezoelectric crystals. Their equivalent circuit is a voltage source in series with a capacitance, as shown in Figure 1-9, which produces an output in coulombs of charge as a function of acceleration excitation. Vibratory acceleration results in an alternating output typically of very small value. Several crystals are therefore stacked to increase the transducer output. As a consequence of the small quantities of charge transferred, this transducer usually is interfaced to a low-input-bias-current charge amplifier, which also converts the acceleration input to a velocity signal. An ac-coupled integrator will then provide a displacement signal that may be calibrated, for example, in millinches of displacement per volt.

A load cell is a transducer whose output is proportional to an applied force. Strain-gauge transducers provide a change in resistance due to mechanical strain produced by a force member. Strain gauges may be based on a thin metal wire, foil, thin films, or semiconductor elements. Adhevise-bonded gauges are the most widely used, with a typical resistive strain element of 350 Ω that will register full-scale changes to 15 Ω. With a Wheatsone-bridge circuit, a 2-V excitation may

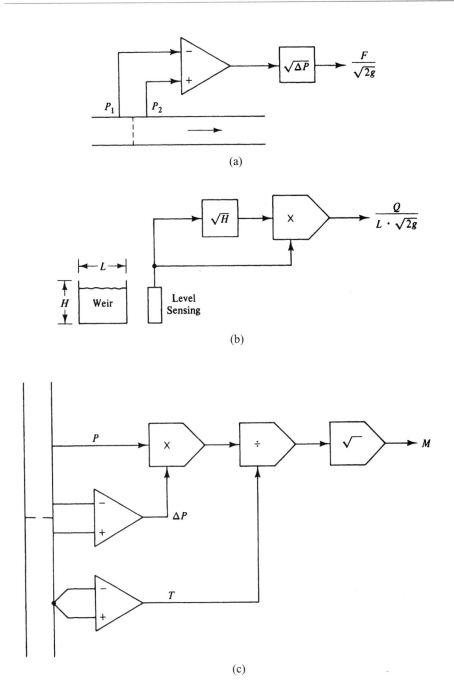

Figure 1-8. (a) Flow rate, (b) volumetric flow, (c) mass flow.

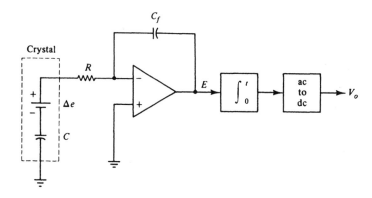

Figure 1-9. Vibration measurement.

therefore provide up to a 50-mV output signal change, as described in Figure 1-10. Semiconductor strain gauges offer high sensitivity at low strain levels, with outputs of 200 mV to 400 mV. Miniature tactile-force sensors can also be fabricated from scaled-down versions of classic transducers by employing MEMS technology. A multiplexed array of these sensors can provide sense feedback for robotic part manipulation and teleoperator actuators.

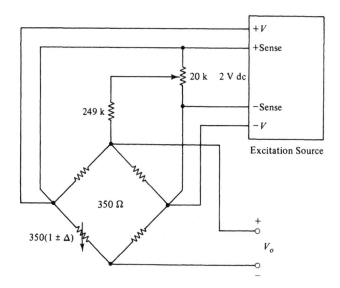

Figure 1-10. Strain gauge.

Ultrasound ranging and imaging systems are increasingly being applied for industrial and medical purposes. A basic ultrasonic system is illustrated by Figure 1-11 consisting of a phased-array transducer and associated signal processing, including aperture focusing by means of time delays, that is employed in both medical ultrasound and industrial nondestructive-testing applications. Multiple frequency emissions in the 1–10 MHz range are typically employed to prevent spatial multipath cancellations. B-scan ultrasonic imaging displays acoustic reflectivity for a focal plane, and C-scan imaging provides integrated volumetric reflectivity of a region around the focal plane.

Hall-effect transducers, which usually are silicon-substrate devices, frequently include an integrated amplifier to provide a high-level output. These devices typically offer an operating range from −40 to +150°C and a linear output. Applications include magnetic-field sensing and position sensing with circuit isolation, such as the Micro Switch LOHET device, which offers a 3.75-mV/gauss response. Figure 1-12 describes the principle of Hall-effect operation. When a magnetic field B_z is applied perpendicular to a current-conducting element, a force acts on the current I_x creating a diversion of its flow proportional to a difference of potential. This measurable voltage V_y is pronounced in materials such as InSb and InAs, and occurs to a useful degree in Si. The magnetic field usually is provided as a function of a measurand.

1-4 QUANTUM SENSORS

Quantum sensors are of significant interest as electromagnetic spectrum transducers over a frequency range extending from the far-infrared region of 10^{11} Hz

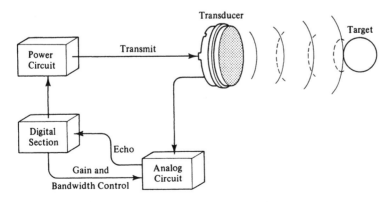

Figure 1-11. Phased-array ultrasound system.

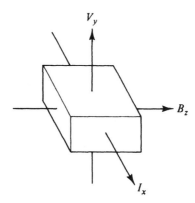

Figure 1-12. Hall-effect transducer.

through the visible spectrum about 10^{14} Hz to the far-ultraviolet region at 10^{17} Hz. These photon sensors are capable of measurements of a single photon whose energy E equals hv, or watt-seconds in radiometry units from Table 1-3, where h is Planck's constant of 6.62×10^{-34} joule-seconds and v is frequency in hertz. Frequencies lower than infrared represent the microwave region and higher than ultraviolet constitute X-rays, which require different transducers for measurement. In photometry, one lumen represents the power flux emitted over one steradian from a source of one candela intensity. For all of these sensors, incident photons result in an electrical signal by an intermediate transduction process.

Table 1-4 describes the relative performance between principal sensors, whereby in photo diodes photons generate electron–hole pairs within the junction depletion region. Photo transistors offer signal gain at the source for this transduction process over the basic photo diode. In photoconductive cells, photons generate carriers that lower the sensor bulk resistance, but their utility is limited

Table 1-3. Quantum sensor units

Parameter	Radiometry	Photometry	Photonic
Energy	watt • sec	lumen • sec	photon
Irradiance	watts/cm^2	footcandles	photon/sec/cm^2
Flux	watts	lumens	photons/sec
Area radiance	$\dfrac{\text{watts / steradian}}{\text{cm}^2}$	footlamberts	photon/sec/cm^2
Point intensity	watts/steradian	candelas • steradian	photon/sec/steradian

Table 1-4. Sensor relative performance

Device	λ Region	$I_{photocurrent}/F_{photons/sec}$	Application
Photo Diode	UV–near IR	10^{-3} amp/watt	Photonic detector
Photoconductive	Visible–near IR	1 amp/watt	Photo controller
Bolometer	Near IR–far IR	10^3 amp/watt	Superconducting IR
Photomultiplier	UV–near IR	10^6 amp/watt	Spectroscopy

by a restricted frequency response. These sensors are shown in Figures 1-13 and 1-14. In all applications, it is essential to match sources and sensors spectrally in order to maximize energy transfer. For diminished photon sources, the photomultiplier excels owing to a photoemissive cathode followed by high multiplicative gain to 10^6 from its cascaded dynode structure. The high gain and inherent low noise provided by coordinated multiplication results in a widely applicable sensor, except for the infrared region. Presently, the photomultiplier vacuum electron ballistic structure does not have a solid-state equivalent.

The measurement of infrared radiation is difficult as a result of the low energy of the infrared photon. This sensitivity deficiency has been overcome by thermally responsive resistive bolometer microsensors employing high-T_c superconductive films, whereby operation is maintained near the film-transition temperature such that small temperature variations from infrared photons provide large resistance changes with gains to 10^3. Microsensor fabrication enhances reliability, and extension to arraying of elements is described by Figure 1-15, with image intensity $\bar{I} = (x,y)$ quantized into a grey-level representation $f(\bar{I})$. This versatile imaging device is employed in applications ranging from analytical spectroscopy to night vision and space defense.

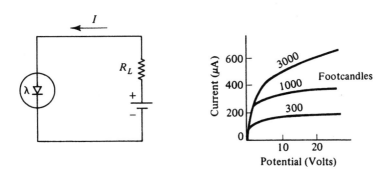

Figure 1-13. Photodiode characteristics.

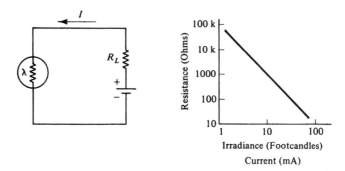

Figure 1-14. Photoconductive characteristics.

A property common to all nuclear radiation is its ability to interact with the atoms that constitute all matter. The nature of the interaction with any form of matter varies with the different components of radiation, as illustrated in Figure 1-16. These components are responsible for interactions with matter that generally produce ionization of the medium through which they pass. This ionization is the principal effect used in the detection of the presence of nuclear radiation. Alpha

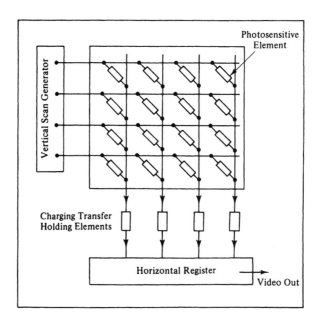

Figure 1-15. Quantum sensor array.

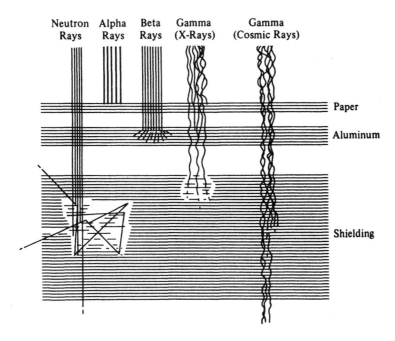

Figure 1-16. Nuclear radiation characteristics.

and beta rays often are not encountered because of their attenuation. Instruments for nuclear radiation detection therefore most commonly constructed to measure gamma radiation and its scintillation or luminescent effect. The rate of ionization in roentgens per hour is a preferred measurement unit, and represents the product of the emanations in curies and the sum of their energies in MeV, represented as gamma energies. A distinction also should be made between disintegrations in counts per minute and ionization rate. The count-rate measurement is useful for half-life determination and nuclear detection, but does not provide exposure rate information for interpretation of degree of hazard. The estimated yearly radiation dose to persons in the United States is 0.25 roentgen (R). A high-radiation area is one in which radiation levels exceed 0.1 R per hour, and requires posting of a caution sign.

Methods for detecting nuclear radiation are based on means for measuring the ionizing effects of these radiations. Mechanizations fall into the two categories of pulse-type detectors of ionizing events, and ionization-current detectors that employ an ionization chamber to provide an averaged radiation effect. The first category includes Geiger–Mueller tubes and more sensitive scintillation counters capable of individual counts. Detecting the individual ionizing scintillations is aided

by an activated crystal such as sodium iodide optically coupled to a high-amplifi-
cation photomultiplier tube, as shown in Figure 1-17. Ionization-current detectors
primarily are employed in health and physics applications such as industrial areas
subject to high radiation levels. An ion chamber is followed by an amplifier
whose output is calibrated in roentgens per hour ionization rate. This method is
necessary where pulse-type detectors are inappropriate because of a very high
rate of ionization events. Practical industrial applications of nuclear radiation and
detection include thickness gauges, nondestructive testing such as X-ray inspec-
tion, and chemical analysis such as by neutron activation.

1-5   ANALYTICAL SENSORS

The relationship between process and analytical measurements often is only a
matter of sensor location, whereby process measurements are acquired during
real-time physical events, whereas analytical measurements may be acquired
post-event offline as an ex-situ assay. These measurements provide useful describing functions that increasingly are applied in a substitutive role for conventional
mathematical-process models. Chemical sensors are employed to determine the
presence, concentration, or quantity of elemental or molecular analytes. These
sensors may be divided into two classes: either electromagnetic, involving filtered
optical and atomic mass unit spectroscopy; or electrochemical, involving the selectivity of charged species. Quantum spectroscopy, described in Figure 1-18, offers specific chemical measurements utilizing wavelength-selective filters from
UV to near-IR coupled to a photoemissive photomultiplier whose output is displayed by a wideband oscilloscope. Alternately, mass-spectrometry chemical
analysis is performed at high vacuum typically employing a quadrupole filter as
shown in Figure 1-19, with sample gas energized by an ion source. The mass filter

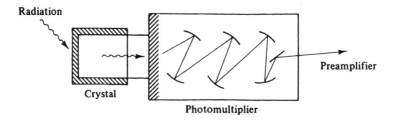

Figure 1-17. Scintillation detector.

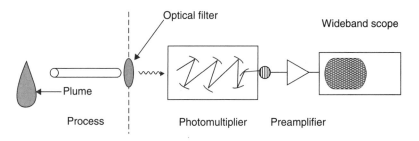

Figure 1-18. Optical spectrometer structure.

selects ions determining specific charge-to-mass ratios, employing both electric and magnetic fields with the relationship $mv^2 = 2\ eV$, that are subsequently collected by an ion detector whose current intensity is displayed versus atomic mass unit (AMU).

Online measurements of industrial processes and chemical streams often require the use of selective chemical analyzers for the control of a processing unit. Examples include oxygen for boiler control, sulfur oxide emissions from combustion processes, and hydrocarbons associated with petroleum refining. Laboratory instruments such as gas chromatographs generally are not used for online measurements primarily because they analyze all compounds present simultaneously rather than a single one of interest.

The dispersive infrared analyzer is the most widely used chemical analyzer, owing to the range of compounds it can be configured to measure. Operation is by the differential absorption of infrared energy in a sample stream in comparison to that of a reference cell. Measurement is by deflection of a diaphragm separating the sample and reference cells, which in turn detunes an oscillator circuit to provide an

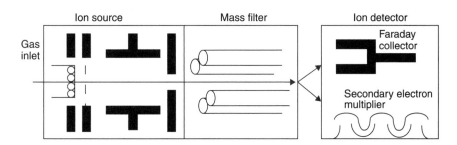

Figure 1-19. Mass spectrometer structure.

electrical analog of compound concentration. Oxygen analyzers usually are of the amperometric type, in which oxygen is chemically reduced at a gold cathode, resulting in a current flow from a silver anode as a function of this reduction in oxygen concentration. In a paramagnetic wind device, a wind effect is generated when a mixture containing oxygen produces a gradient in a magnetic field. Measurement is derived by the thermal cooling effect on a heated resistance element forming a thermal anemometer. Table 1-5 describes basic electrochemical analyzer methods, and Figure 1-20 a basic gas-analyzer system with calibration.

Also in this group are pH, conductivity, and ion-selective electrodes. pH defines the balance between the hydrogen ions H^+ of an acid and the hydroxyl ions OH^- of an alkali; one type can be increased only at the expense of another. A pH probe is sensitive to the presence of H+ ions in solution, thereby representing the acidity or alkalinity of a sample. All of these ion-selective electrodes are based on the Nernst equation (1-7), which typically provides a 60-mV potential change for each tenfold change in the activity of a monovalent ion.

$$V_0 = V + \frac{F}{n} \log(ac + s_1 a_1 c_1 + ...) \text{ volts} \qquad (1-7)$$

V_0 = voltage between sensing and reference electrodes

V = electrode base potential

F = Nernst factor, 60 mV at 25°

n = ionic charge, 1 monovalent, 2 bivalent, etc.

a = ionic activity

c = concentration

s = electrode sensitivity to interfering ions

Online near-infrared sensors are typically employed for process and quality measurements, primarily for the control of content and film thickness of water and

Table 1-5. Chemical analyzer methods

Compound	Analyzer
CO, SO_x, NH_x	Infrared
O_2	Amperometric, paramagnetic
HC	Flame ionization
NO_x	Chemiluminescent
H_2S	Electrochemical cell

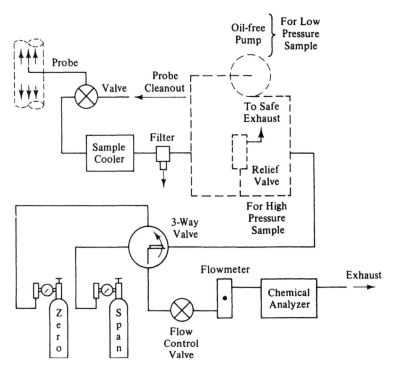

Figure 1-20. Calibrated gas analyzer.

organic materials, contained by products from paper to polymers to steel. Measurement error to 0.1%FS is available with sample illumination, absorption, and reflectance at NIR wavelengths between 1100 and 2500 nm. The near-infrared analytical sensor in Figure 1-21 describes a mechanization whereby different infrared filters are employed for measurement (absorption wavelength) and reference (reflectance wavelength) to achieve a ratio that compensates for variabilities encountered in the application environment. Sensor calibration is achieved by means of span and zero that, respectively, adjust sensitivity and offset.

The concept of computer-based instruments arose with the advent of inexpensive computation, furthered by the personal computer, which permitted networking discrete instruments into sophisticated automated test systems beginning in the 1970s. The evolution of more efficient data acquisition and presentation, resulting from user-defined programmability and reconfigurability, continues through the present to provide a more computationally intensive instrumentation framework. Contemporary virtual instruments accordingly are capable of elevating fundamental sensor data to a substantially higher attribution, enabling more complex cogni-

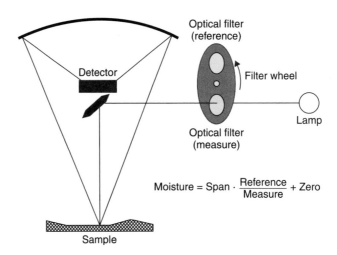

$$\text{Moisture} = \text{Span} \cdot \frac{\text{Reference}}{\text{Measure}} + \text{Zero}$$

Figure 1-21. Near-infrared analytical sensor. (Courtesy Moisture Systems Corporation.)

tive interpretation. Multifunction I/O hardware is typically combined with application-development software on a personal computer platform for the realization of specific virtual instruments such as the following microwave microscopy example for sample essays in manufacturing and biomedical applications.

A benefit of microwave microscopy is micron-resolution sample imaging of subsurface as well as surface properties. With microwave excitation wavelengths on the order of millimeters, their limiting half-wavelength Abbe resolution barrier is extended through detection of shorter near-field evanescent microwave spatial wavelength components aided by enhanced transducer excitation, sensitivity, and signal processing. With the virtual instrument of Figure 1-22, a 100-Hz sinusoidal 1F signal dither of the 30 GHz microwave source enables synthesis of 2F spatial frequency components associated with a sample. Sensitivity is increased by an automatic controller whose dc controlled variable tunes the microwave source, as does the dither signal, to obtain a resonant frequency shift that maximizes the spatial components. The greater the 2F signal DFT-normalized magnitude achieved, the more selective the resonant curve quality factor and frequency shift images become relative to the sample. The resolution of this instrument is limited by −55dB of spurious noise from contemporary GHz sources, which is equivalent to nine-bit accuracy by Table 5-7 (see Chapter 5).

The architecture of virtual instrument software may be divided into two layers: measurement and configuration services, and application development tools. Measurement and configuration services contain prescriptive software drivers for

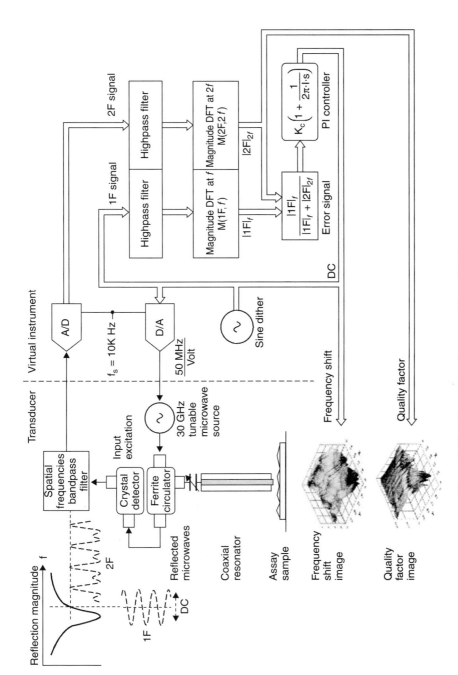

Figure 1-22. Microwave microscopy biomedical instrument.

23

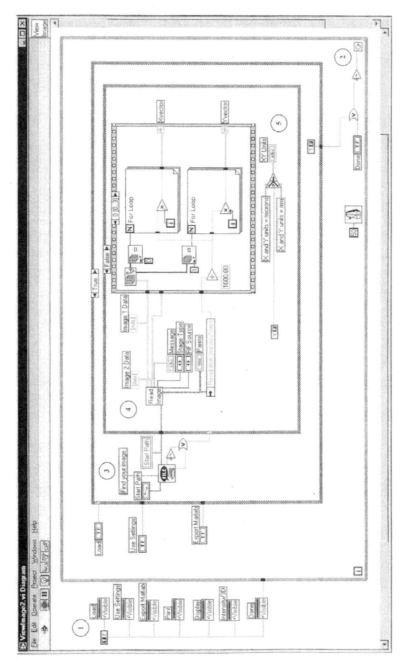

Figure 1-23. LabVIEW display generation graphical program.

24

interfacing hardware I/O devices as subroutines that are usually accessed by graphical icons. Configuration utilities are also included in this layer for naming and setting hardware channel attributes such as amplitude scaling. Software selected for application development may be sourced separately from hardware devices only when compatibility is ensured.

Graphical programs typically consist of an icon diagram including a front panel that serves as the source code for an application program. The front panel provides a graphical user interface for functions to be executed concurrently. The Lab-VIEW diagram of Figure 1-23 shows a view image module for generating the data display images shown in Figure 1-22, which may be exported as Matlab files. When this program is initiated, the front panel (1) defines display visibility attributes. Assets within a "while loop" are then executed cyclically until control Done is set false (2), allowing conditional expressions to break this "while loop." The metronome icon describes a 50-millisecond interval within which the "while loop" iterates. The data structure that performs this image generation executes sequentially.

The concentric window (3) is analogous to a "case" statement in the C language controlled by the Boolean Load variable. The "case" procedure occurs when Load is true. The icons within (3) allow a user to locate an image file, whereas the icons within (4) provide a subprogram for extracting microscopy image content for arrays representing frequency shift and quality factor data. A "sequence" data structure performs total image formatting. Icons within (4) select strings contained in (5) to write data images as XY vectors. Note that this code is set to display X and Y in millimeter units.

PROBLEMS

1-1. Sensor understanding is an essential requirement for design and specification. Compare the instrumentation application features of thermocouples, RTDs, and semiconductor temperature sensors.

1-2. A potentiometric transducer device contains 1800 wire turns totaling 1K ohms resistance. Determine the output noise voltage amplitude generated for 1V dc excitation per wire turn contact movement.

1-3. Perform a linearizer design for a Type-S thermocouple using quadratic linearization with coefficient solution at 300°C and 600°C. Intrinsic thermocouple output is described by the following table.

°C	0	150	300	450	600	750
mV	0.016	1.03	2.32	3.74	5.23	6.80

1-4. A radiometer weather satellite sensor, with available mean time between failure (MTBF) data in hours, permits determining the reliability for continuous operating intervals R_{TOT} by predicting survival for random subsystem failures (Figure 1-24). Evaluate satellite reliability for a five-year operating time interval t.

$$R_{TOT} = R_{ANT} \times R_{REC} \times R_{MOD} \times R_{PWR}$$

where

$$R = \exp(-t/MTBF)$$

Radiometer element	MTBF (hrs)
Antenna	9×10^5
Receiver	5×10^5
MODEM	7×10^5
Power	3×10^5

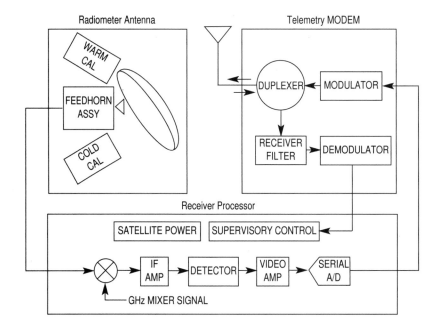

Figure 1-24.

BIBLIOGRAPHY

1. Gardner, J.W., *Microsensors,* Wiley, 1994.
2. Garrett, P.H., et al., *Advanced Instrumentation and Computer I/O Design,* IEEE Press, 1994.
3. Garrett, P.H., et al., "Emerging Methods for the Intelligent Processing of Materials," *J. of Materials Engineering and Performance,* Vol. 2, No. 5, October 1993.
4. Garrett, P.H. (contributing author), *Handbook of Industrial Automation,* Marcel Dekker, 2000.
5. Garrett, P.H., et al., "Self-Directed Processing of Materials," IFAC, *Engineering Applications of Artificial Intelligence,* Elsevier, Vol. 12, August 1999.
6. Jha, A.R., *Infrared Technology,* Wiley, 2000.
7. Kovacs, G.T.A., *Micromachined Transducers Sourcebook,* McGraw-Hill, 1998.
8. Malas, J.C., et al., "Emerging Sensors for the Intelligent Processing of Materials," *Journal of Materials,* Vol. 48, No. 9, September 1996.
9. Norton, H.N., *Handbook of Transducers for Electronic Measuring Systems,* Prentice-Hall, 1969.
10. Petriu, E.M. (Editor), *Instrumentation and Measurement Technology and Applications,* IEEE Technology Update Series, 1998.
11. Prensky, S.D. and Castellucis, R.L., *Electronic Instrumentation,* 3rd ed., Prentice-Hall, 1982.
12. Rabinovich, S.G., *Measurement Errors and Uncertainties,* 2nd ed., Springer-Verlag, 1999.
13. Tabib-Azar, M., et al., "Super-Resolution Characterization of Microwave Conductivity of Semiconductors," *IOP Measurement Science Techn.,* Vol. 3, 1993.
14. Draper, C.S., *Instrument Engineering, Vol. 1, Fundamentals,* McGraw-Hill, 1952.

CHAPTER 2

Instrumentation Amplifiers and Parameter Errors

2-0 INTRODUCTION

This chapter is concerned with the devices and circuits that comprise the electronic amplifiers of linear systems utilized in instrumentation applications. This development begins with the temperature limitations of semiconductor devices, which is then extended to differential amplifiers and an analysis of their parameters for understanding operational amplifiers from the perspective of their internal stages. This includes gain–bandwidth–phase stability relationships and interactions in multiple amplifier systems. An understanding of the capabilities and limitations of operational amplifiers is a prerequisite to understanding instrumentation amplifiers.

An instrumentation amplifier usually is the first electronic device encountered in a signal acquisition system, and in large part it is responsible for the ultimate data accuracy attainable. Present instrumentation amplifiers are shown to possess sufficient linearity, common-mode rejection ratio (*CMRR*), low noise, and precision for total errors in the microvolt range. Five categories of instrumentation amplifier applications are described with representative contemporary devices and parameters provided for each. These parameters are then utilized to compare amplifier circuits for implementations ranging from low input voltage error to wide bandwidth applications.

Advanced Instrumentation and Computer I/O Design, Second Edition. By Patrick H. Garrett
Copyright © 2013 the Institute of Electrical and Electronics Engineers, Inc.

2-1   DEVICE TEMPERATURE CHARACTERISTICS

The elemental semiconductor device in electronic circuits is the pn junction; among its forms are diodes and bipolar and FET transistors. The availability of free carriers that result in current flow in a semiconductor is a direct function of the applied thermal energy. At room temperature, taken as 20°C (293°K above absolute zero), there is abundant energy to liberate the valence electrons of a semiconductor. These carriers are then free to drift under the influence of an applied potential. The magnitude of this current flow is essentially a function of the thermal energy instead of the applied voltage and accounts for the negative temperature coefficient exhibited by semiconductor devices (increasing current with increasing temperature).

The primary variation associated with reverse-biased pn junctions is the change in reverse saturation current I_s with temperature. I_s is determined by device geometry and doping with a variation of 7% per degree centigrade both in silicon and germanium, doubling with every 10°C rise. This behavior is shown by Figure 2-1 and Equation (2-1). Forward-biased pn junctions exhibit a decreasing junction potential, having an expected value of –2.0 mV per degree centigrade rise, as defined by Equation (2-2). The dV/dT temperature variation is shown to be the difference between the forward-junction potential V and the temperature of I_s. This relationship is the source of the voltage offset drift with temperature exhibited by semiconductor devices. The volt-equivalent of temperature is an empirical model in both equations, defined as $V_T = (273°K + T°C)/11{,}600$, having an expected value of 25 mV at room temperature.

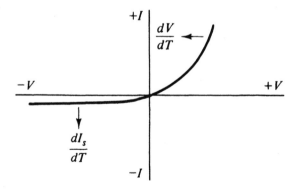

Figure 2-1. pn Junction temperature dependence.

$$\frac{dI_s}{dT} = I_s \cdot \frac{d(\ln I_s)}{dT} \; \text{A/°C} \qquad \text{Reversed biased} \qquad (2\text{-}1)$$

$$\frac{dV}{dT} = \frac{V}{T} - V_T \left(\frac{1}{I_s} \cdot \frac{dI_s}{dT} \right) \text{V/°C} \qquad \text{Forward biased} \qquad (2\text{-}2)$$

2-2 DIFFERENTIAL AMPLIFIERS

The first electronic circuit encountered by a sensor signal in a data acquisition system typically is the differential input stage of an instrumentation amplifier. The balanced bipolar differential amplifier of Figure 2-2(a) is an important circuit used in many linear applications. Operation with symmetrical plus minus power supplies as shown results in the input base terminals being at 0 V under quiescent conditions. Due to the interaction that occurs in this emitter-coupled circuit, the algebraic difference signal applied across the input terminals is the effective drive signal, whereas equally applied input signals are cancelled by the symmetry of the circuit. With reference to a single-ended output V_{o2}, amplifier Q_1 may be considered an emitter follower with the constant-current source an emitter load impedance in the megohm range. This results in a noninverting voltage gain for Q_1 very close to unity (0.99999), that is, emitter coupled to the common-emitter amplifer Q_2, where Q_2 provides the differential voltage gain A_{vdiff} by Equation (2-3).

Differential amplifier volt-ampere transfer curves are defined by Figure 2-2(b), where the abscissa represents normalized differential input voltage $(V_1 - V_2)/V_T$. The transfer characteristics are shown to be linear about the operating point corresponding to an input-voltage swing of approximately 50 mV ($\pm$ 1 V_T unit). The maximum slope of the curves occurs at the operating point of $I_o/2$, and defines the effective transconductance of the circuit as $\Delta I_C/\Delta(V_1 - V_2)/V_T$. The value of this slope is determined by the total current I_o of Equation (2-4). Differential input impedances R_{diff} and R_{cm} are defined by Equations (2-5) and (2-6). The effective voltage gain cancellation between the noninverting and inverting inputs is represented by the common-mode gain A_{vcm} of Equation (2-7). The ratio of differential gain to common-mode gain also provides a dimensionless figure of merit for differential amplifiers as the common-mode rejection ratio (*CMRR*). This is expressed by Equation (2-8), having a typical value of 10^5.

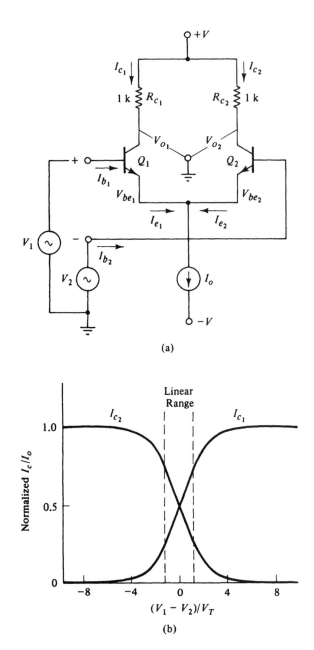

(a)

(b)

Figure 2-2. Differential dc amplifier and transfer curves. $h_{fe} = 100$, $h_{ie} = 1k$, $h_{oe} = 10^{-6}$ mho.

$$A_{v_{diff}} = \frac{h_{fe}R_c}{2h_{ie}} \quad \text{single-ended } V_{o_2} \tag{2-3}$$

$$= 50$$

$$I_o = I_{s1} \cdot \exp\left(V_{be1}/V_T\right) + I_{s2} \cdot \exp\left(V_{be2}/V_T\right) \tag{2-4}$$

$$= 1 \text{ mA}$$

$$R_{diff} = \frac{4V_T h_{fe}}{I_o} \tag{2-5}$$

$$= 10 \text{ K}$$

$$R_{cm} = \frac{h_{fe}}{h_{oe}} \tag{2-6}$$

$$= 100 \text{ M}$$

$$A_{v_{cm}} = \frac{h_{oe}R_c}{2} \tag{2-7}$$

$$= 5 \times 10^{-4}$$

$$CMRR = \frac{A_{v_{diff}}}{A_{v_{cm}}} \tag{2-8}$$

$$= 10^5$$

The performance of operational and instrumentation amplifiers is largely determined by the errors associated with their input stages. It is convention to express these errors as voltage and current offset values, including their variation with temperature with respect to the input terminals, so that various amplifiers may be compared on the same basis. In this manner, factors such as the choice of gain and the amplification of the error values do not result in confusion concerning their true magnitude. It is also notable that the symmetry provided by the differential amplifier circuit primarily serves to offer excellent dc stability and the minimization of input errors in comparison with those of nondifferential amplifiers.

The base-emitter voltages of a group of the same type of bipolar transistors at the same collector current are typically only within 20 mV. Operation of a differential pair with a constant-current emitter sink as shown in Figure 2-2(a), however, provides a V_{be} match of V_{os} of about 1 mV. Equation (2-9) defines this input offset voltage and its dependence on the mismatch in reverse saturation current I_s between the differential pair. This mismatch is a consequence of variations in doping and geometry of the devices during their manufacture. Offset adjustment is frequently provided by the introduction of an external trimpot R_{Vos} in the emitter circuit shown in Figure 2-4. That permits the incremental addition and subtraction of emitter voltage to drop to 0 V_{os} without disturbing the emitter current I_o.

Of greater concern is the offset voltage drift with temperature dV_{os}/dT. This input error results from mistracking of V_{be1} and V_{be2}, described by Equation (2-10), and is difficult to compensate. However, the differential circuit reduces dV_{os}/dT to $2\mu V/°C$ from the $-2mV/°C$ for a single device of Equation (2-2), for an improvement factor of 1/1000. By way of comparison, JFET differential circuit V_{os} is larger and on the order of 10 mV, and dV_{os}/dT typically $5\mu V/°C$. Minimization of these errors is achieved by matching the device pinch-off voltage parameter. Bipolar input bias-current offset and offset-current drift are described by Equations (2-11) and (2-12), and have their genesis in a mismatch in current gain ($h_{fe1} \neq h_{fe2}$). JFET devices intrinsically offer lower input bias currents and offset-current errors in differential circuits, which is advantageous for the amplification of current-type sensor signals. However, the rate of increase of JFET bias current with temperature is exponential, as illustrated in Figure 2-3, and results in values that exceed bipolar input bias currents at temperatures beyond 100°C, thereby limiting the utility of JFET differential amplifiers above this temperature.

$$V_{os} = V_T \ln \frac{I_{s2}}{I_{s1}} \cdot \frac{I_{e1}}{I_{e2}} \qquad (2\text{-}9)$$

$$= 1 \text{ mV}$$

$$\frac{dV_{os}}{dT} = \frac{dV_{be1}}{dT} - \frac{dV_{be2}}{dT} \qquad (2\text{-}10)$$

$$= 2 \ \mu V/°C$$

$$I_{os} = I_{b1} - I_{b2} \qquad (2\text{-}11)$$

$$= 50 \text{ nA}$$

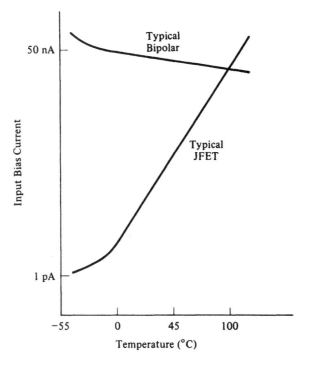

Figure 2-3. Transistor input current temperature drift.

$$\frac{dI_{os}}{dT} = B \cdot I_{os} \tag{2-12}$$

$$= 0.25 \text{ nA/}°\text{C}$$

$$B = -0.005/°\text{C} > 25°\text{C}$$

$$= -0.015/°\text{C} < 25°\text{C}$$

2-3   OPERATIONAL AMPLIFIERS

Most operational amplifiers are of similar design, as described by Figure 2-4, and consist of a differential-input stage cascaded with a high-gain interstage followed by a power-output stage. Operational amplifiers are characterized by very high gain at dc and a uniform rolloff in this gain with frequency. This enables these de-

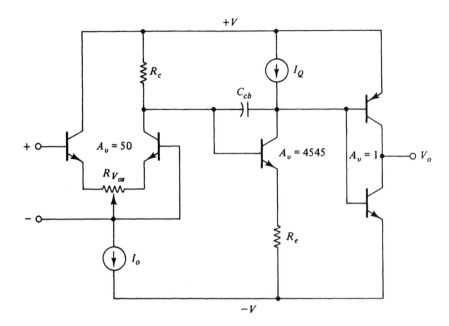

Figure 2-4. Elemental operational amplifier circuit.

vices to accept feedback from arbitrary networks with high stability and simultaneous dc and ac amplification. Consequently, such networks can accurately impart their characteristics to electronic systems with negligible degradation. The earliest integrated-circuit amplifier was offered in 1963 by Texas Instruments, but the Fairchild 709 introduced in 1965 was the first operational amplifier to achieve widespread application. Improvements in design resulted in second-generation devices such as the National LM108. Advances in fabrication technology enabled amplifiers exemplified by the Analog Devices OP-07 with improved performance overall. Subsequent refinements are represented by devices including the Linear LTC-1250 featuring zero drift and ultralow noise. It is notable that contemporary operational amplifier circuits are structured around a high-gain innerstage employing a constant-current source active load. The gain stage active load impedance of approximately 500 kohms ratioed with an emitter resistance R_e approximating 100 ohms, shown in Figure 2-4, is responsible for high overall A_{V_o}.

The circuit for an inverting operational amplifier is shown in Figure 2-5. The cascaded innerstage gains of Figure 2-4 provide a total open-loop gain A_{V_o} of 227,500, enabling realization of the ideal closed-loop gain A_{V_c} representation of Equation (2-13).

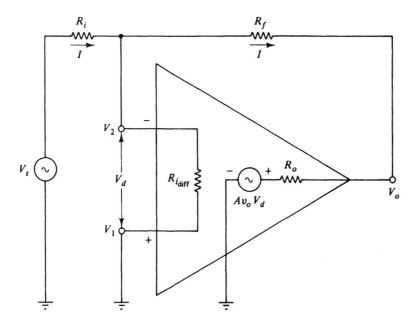

Figure 2-5. Inverting operational amplifier.

$$A_{V_c} = \frac{V_o}{V_s} = \frac{-IR_f}{IR_c} = \frac{-R_f}{R_i} \qquad (2\text{-}13)$$

since

$$R_{i_{diff}} \rightarrow \infty, \ Vd = \frac{V_o}{A_{V_o}} \rightarrow 0 \ as \ |A_{V_o}| \rightarrow \infty$$

In practice, the A_{V_o} value cannot be utilized without feedback because of nonlinearities and instability. The introduction of negative feedback between the output and inverting input also results in a virtual ground with equilibrium current conditions maintaining $V_d = V_1 - V_2$ at zero. Classification of operational amplifiers is primarily determined by the active devices that implement the amplifier differential input. Table 2-1 delineates this classification.

According to negative-feedback theory, an inverting amplifier will be unstable if its gain is equal to or greater than unity when the phase shift reaches $-180°$ through the amplifier. This is so because an output-to-input relationship will also have been established, providing an additional $-180°$ by the feedback network. The relationships between amplifier gain, bandwidth, and phase are described by Fig-

Table 2-1. Operational and instrumentation amplifier types

Bipolar	Prevalent type used for a wide range of signal-processing applications. Good balance of performance characteristics.
FET	Very high input impedance frequently employed as an instrumentation-amplifier preamplifier. Exhibits larger input errors than bipolar devices.
CAZ	Bipolar device with auto-zero circuitry for internally measuring and correcting input error voltages. Provides low-input-uncertainty amplification.
BiFET	Combined bipolar and FET circuit for extended performance. Intended to displace bipolar devices in general-purpose applications.
Superbeta	A bipolar device approaching FET input impedance with lower bipolar errors. A disadvantage is lack of device ruggedness.
Micropower	High-performance operation down to 1-volt supply powered from residual system potentials. Employs complicated low-power circuit equivalents for implementation.
Isolation	An internal barrier device using modulation or optical methods for very high isolation. Medical and industrial applications.
Chopper	dc-ac-dc circuit with a capacitor-coupled internal amplifier providing very low offset errors for minimum input uncertainty.
Varactor	Varactor diode input device with very low input bias currents for current-amplification applications such as photomultipliers.
Vibrating capacitor	A special input circuit arrangement requiring ultralow input bias currents for applications such as electrometers.

ure 2-6 and Equations (2-14) through (2-16) for an example closed-loop gain A_{V_c} value of 100. Each discrete innerstage contributes a total of $-90°$ to the cumulative phase shift ϕ_t, with $-45°$ realized at the respective -3 dB frequencies. The high-gain stage -3 dB frequency of 10 Hz is attributable to the dominant-pole compensating capacitance C_{cb} shown in Figure 2-4. The second corner frequency at 1 MHz is typical for a differential input stage, and the third at 25 MHz is contributed by the output stage.

$$A_{V_o} = \frac{227.250}{\left(1 + j\dfrac{f}{10 \text{ Hz}}\right)\left(1 + j\dfrac{f}{1 \text{ MHz}}\right)\left(1 + j\dfrac{f}{25 \text{ MHz}}\right)} \qquad (2\text{-}14)$$

$$\phi_t = -\tan^{-1}\left(\frac{f}{10 \text{ Hz}}\right) - \tan^{-1}\left(\frac{f}{1 \text{ MHz}}\right) - \tan^{-1}\left(\frac{f}{25 \text{ MHz}}\right) \qquad (2\text{-}15)$$

$$\text{Phase margin} = 180° - \phi_t \qquad (2\text{-}16)$$

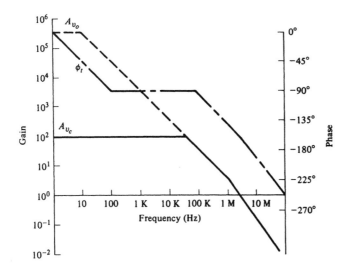

Figure 2-6. Operational amplifier gain–bandwidth-phase relationships.

2-4 INSTRUMENTATION AMPLIFIERS

The acquisition of accurate measurement signals, especially low-level signals in the presence of interference, requires amplifier performance beyond the typical signal-conditioning capabilities of operational amplifiers. An instrumentation amplifier is usually the first electronic device encountered by a sensor in a signal-acquisition channel, and in large part it is responsible for the ultimate data accuracy attainable. Present instrumentation amplifiers possess sufficient linearity, stability, and low noise for total error in the microvolt range even when subjected to temperature variations, and are on the order of the nominal thermocouple effects exhibited by input lead connections. High *CMRR* is essential for achieving the amplifier performance of interest with regard to interference rejection, and for establishing a signal ground reference at the amplifier that can accommodate the presence of ground-return potential differences. High amplifier-input impedance is also necessary to preclude input signal loading and voltage divider effects from finite source impedances, and to accommodate source-impedance imbalances without degrading *CMRR*. The precision gain values possible with instrumentation amplifiers, such as 1000.000, are equally important to obtain accurate scaling and registration of measurement signals.

   The relationship of *CMRR* to the output signal V_o for an operational or instrumentation amplifier is described by Equation (2-17), and is based upon the derivation of *CMRR* provided by Equation (2-8). For the operational amplifier sub-tractor circuit of Figure 2-7, A_{vdiff} is determined by the feedback-to-input resistor ratios (R_f/R_i) with practically realizable values to 100, and A_{vcm} is determined by the mismatch between feedback and input resistor values attributable to their tolerances. Consequently, the A_{vcm} for a subtractor circuit may be obtained from Equation (2-18) and as tabulated in Table 2-2 to determine the average expected *CMRR* value for specified resistor tolerances. Notice that *CMRR* increases with A_{vdiff} by the numerator of Equation (2-8), but A_{vcm} is constant because of its normalization by the resistor tolerance chosen.

$$V_o = A_{vdiff} \cdot V_{diff} + A_{vcm} \cdot V_{cm} \tag{2-17}$$

$$= \left(1 + \frac{1}{CMRR} \cdot \frac{V_{cm}}{V_{diff}} \right)$$

$$CMRR_{\text{subtractor}} = \frac{\dfrac{1}{2} \left(\left| \dfrac{R_{f2} \pm \Delta R_{f2}}{R_{i2} \pm \Delta R_{i2}} \right| + \left| \dfrac{R_{f1} \pm \Delta R_{f1}}{R_{i1} \pm \Delta R_{i1}} \right| \right)}{\left| \dfrac{R_{f2} \pm \Delta R_{f2}}{R_{i2} \pm \Delta R_{i2}} \right| - \left| \dfrac{R_{f1} \pm \Delta R_{f1}}{R_{i1} \pm \Delta R_{i1}} \right|} \tag{2-18}$$

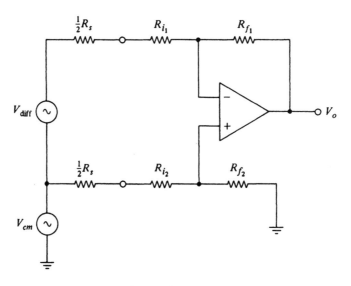

Figure 2-7. Subtractor instrumentation amplifier.

Table 2-2. Subtractor *CMRR* expected values

Resistor tolerance	5%	2%	1%	0.1%
A_{vcm} subtractor	0.1	0.04	0.02	0.002
$CMRR_{subtractor}$ (xA_{vdiff})	10	25	50	500

The subtractor circuit is capable of typical values of *CMRR* to 10^4, and its implementation is economical owing to the requirement for a single operational amplifier. However, its specifications are usually marginal when compared with the requirements of typical signal acquisition applications. For example, each implementation requires the matching of four resistors, and the input impedance is constrained to the value of R_i chosen. For modern bipolar amplifiers, such as the Analog Devices OP-07 and Burr Brown OPA-128 devices with gigohm internal resistances, megohm R_i values are allowable without input voltage divider effects resulting from an imbalanced kilohm R_s source resistance. Low-bias-current amplifiers are essential for current sensors, including nuclear gauges, pH probes, and photomultiplier tubes. The OPA-128 offers a balance of input parameters for this application with an I_{os} of 30 fA and typical current sensor R_s values of 10 M ohms. The compensating resistor R_c shown in Figure 2-8 is matched to R_s in order to preserve *CMRR*. The five amplifiers presented in Table 2-3 beneficially

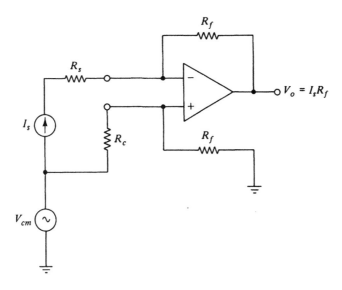

Figure 2-8. Differential current–voltage amplifier.

Table 2-3. Amplifier input parameters define interface applications

Parameter	Low offset voltage OP-07	Low bias current OPA-128	Three-amplifier instrumentation AD624	High-voltage isolation AD215	Wideband video OPA-646	Comment
V_{OS}	10 µV	140 µV	25 µV	0.4mV	1mV	Offset voltage
$\dfrac{dV_{OS}}{dT}$	0.2 µV/°C	5 µV/°C	0.25 µV/°C	2 µV/°C	12 µV/°C	Offset voltage drift
I_{OS}	0.3 nA	30 fA	10 nA	300 nA	0.4 µA	Offset current
$\dfrac{dI_{OS}}{dT}$	5 pA/°C	Negligible	20 pA/°C	1 nA/°C	10 nA/°C	Offset current drift
Sr	0.3 V/µs	3 V/µs	5 V/µs	6 V/µs	180 V/µs	Slew rate
f_{hi}	600 KHz	500 KHz	1 MHz	120 KHz	650 MHz	Unity gain bandwidth
CMRR	10^4	10^4	10^5	10^5	10^4	A_{vdiff}/A_{vcm}
V_{cm}	10 Vrms	10 Vrms	10 Vrms	1500 Vrms	10 Vrms	Maximum applied volts
V_n rms	10 nV/$\sqrt{\text{Hz}}$	27 nV/$\sqrt{\text{Hz}}$	4 nV/$\sqrt{\text{Hz}}$	Negligible	7.5 nV/$\sqrt{\text{Hz}}$	Voltage noise
$f(Av)$	0.01%	0.01%	0.001%	0.0005%	0.025%	Gain nonlinearity
$\dfrac{dAv}{dT}$	50 ppm/°C	50 ppm/°C	5 ppm/°C	15 ppm/°C	50 ppm/°C	Gain drift
R_{diff}	$8 \times 10^7\ \Omega$	$10^{13}\ \Omega$	$10^9\ \Omega$	$10^{12}\ \Omega$	15 KΩ	Differential resistance
R_{cm}	$2 \times 10^{11}\ \Omega$	$10^{13}\ \Omega$	$10^9\ \Omega$	$5 \times 10^9\ \Omega$	1.6 MΩ	Common-mode resistance

permit the comparison of parameters that influence performance in specific amplifier applications, where the *CMRR* entries described are expected in-circuit values.

The three-amplifier instrumentation amplifier of Figure 2-9, exemplified by the AD624, offers improved performance overall compared to the foregoing subtractor circuit with in-circuit $CMRR_{3ampl}$ values of 10^5 and the absence of problematic external discrete input resistors. In order to minimize output noise and offsets with this amplifier, its subtractor A_{vdiff} is normally set to unity gain. The first stage of this amplifier also has a unity A_{vcm}, owing to its differential-input-to-differential-output connection, which results in identical first-stage *CMRR* and A_{vdiff} values. Amplifier internal resistance trimming consequently achieves the nominal subtractor A_{vcm} value shown in Equation (2-19).

$$CMRR_{3\ ampl} = CMRR_{1st\ stage} \cdot CMRR_{subtractor} \qquad (2\text{-}19)$$

$$= \frac{A_{v_{diff\ 1st\ stage}}}{1} \cdot \frac{1}{A_{v_{cm\ subtractor}}}$$

$$= \left(1 + \frac{2R_0}{R_1}\right) \cdot \left(\frac{1}{0.001}\right)$$

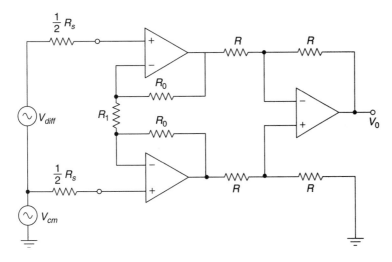

Figure 2-9. Three-amplifier instrumentation amplifier.

The differential output instrumentation amplifier, illustrated by Figure 2-10, offers increased common-mode rejection via Equation (2-20) over the three-amplifier circuit from the addition of a second output subtractor. By comparison, a single subtractor permits a full-scale 24 V_{pp} output signal swing while dual subractors deliver a full-scale 48 V_{pp} output signal from opposite polarity swings of the ±15 V dc power supplies for each signal half-cycle. The effective output gain doubling combined with first-stage gain provides $CMRR_{diff}$ output values to 10^6. This advanced amplifier circuit permits high-performance analog signal acquisition and the continuation of common-mode interference rejection over a signal transmission channel, with termination by a remote differential-to-single-ended subtractor amplifier.

$$CMRR_{diff\ output} = \left(1 + \frac{2R_0}{R_G}\right) \cdot \left(\frac{2}{0.001}\right) \qquad (2\text{-}20)$$

Isolation amplifiers are advantageous for very noisy and high-voltage environments plus the interruption of ground loops. They further provide galvanic isolation typically on the order of 1 μA input-to-output leakage. The front end of the iso amplifier is similar to an instrumentation amplifier, as shown in Figure 2-11, and is operated from an internal dc–dc isolated power supply to insure isola-

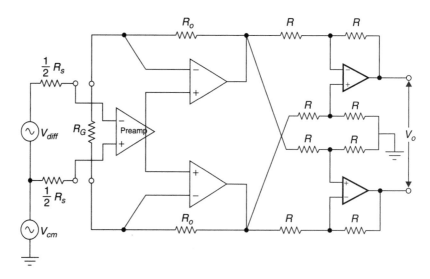

Figure 2-10. Differential output instrumentation amplifier.

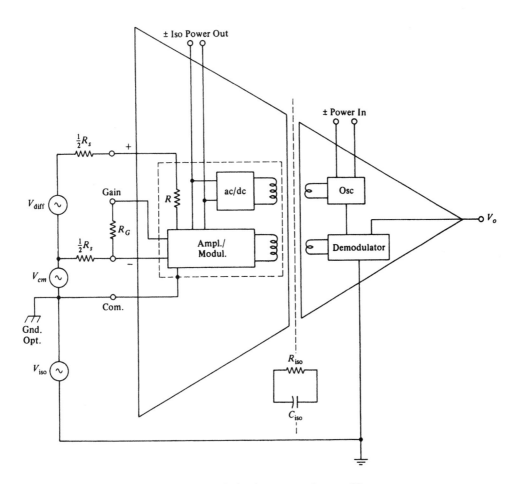

Figure 2-11. Isolation instrumentation amplifier.

tion integrity and for external sensor excitation purposes. As a consequence, these amplifiers do not require sourcing or sinking external bias current, and function normally with fully floating sensors. Most designs also include a 100 Kohm series input resistor R to limit catastrophic fault currents. Typical isolation barriers have an equivalent circuit of 10^{11} ohms shunted by 10 pF representing R_{iso} and C_{iso}. An input-to-output V_{iso} rating of 1500 Vrms is common, and has a corollary isolation mode rejection ratio (IMRR) with reference to the output. *CMRR* values of 10^5 relative to the input common, and IMRR values to 10^8 with reference to the output, are available at 60 Hz. This capability makes possible the accommodation of two sources of interference, V_{cm} and V_{iso}, both frequently encountered in sensor applications. The performance of this connection is described by Equation (2-21).

$$V_o = A_{v_{diff}} \cdot V_{diff}\left[1 + \frac{1}{CMRR} \cdot \frac{V_{cm}}{V_{diff}}\right] + \frac{V_{iso}}{IMRR} \tag{2-21}$$

High-speed data conversion and signal conditioning circuits capable of accommodating pulse and video signals require wideband operational amplifiers. Such amplifiers are characterized by their settling time, delay, slew rate, and transient subsidence, described in Figure 2-12. Parasitic reactive circuit elements and carelessly planned circuit layouts can result in performance derogation. Amplifier slew rate depends directly upon the product of the output voltage amplitude and signal frequency, and this product cannot exceed the slew-rate specification of an amplifier if linear performance is to be realized. For example, a 1 V_{pp} sine wave signal at a frequency of 3 MHz typically encountered in video systems specifies an amplifier slew rate of at least 9.45 V/μs. If the amplifier is also loaded by 1000 pF of capacitance, then it must also be capable of delivering 10 mA of current output at that frequency. These relationships are described by Equation (2-22) and its nomograph of Figure 2-13.

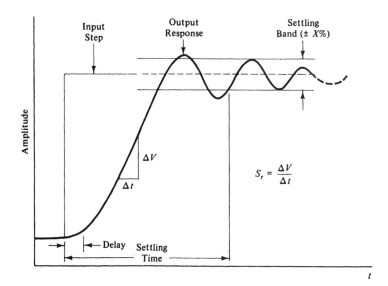

Figure 2-12. Wideband-amplifier settling characteristics.

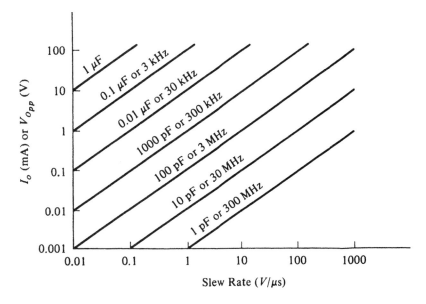

Figure 2-13. Amplifier slew-rate curves.

$$S_r = V_{o_{pp}} \cdot \pi \cdot f_{\text{signal}} \qquad (2\text{-}22)$$

$$= \frac{I_o}{C_{sh}} \text{ V/s}$$

2-5   AMPLIFIER PARAMETER ERROR EVALUATION

The selection of an instrumentation amplifier involves the choice of amplifier input parameters that minimize amplification errors for applications of interest. It is therefore instructive to perform an error comparison between the five diverse amplifier types listed in Table 2-3, considering application-specific V_{cm} and R_s input values, with evaluation of voltage offsets, interference rejection, and gain nonlinearity. The individual error totals tabulated in Table 2-4 provide a performance summary expressed both as a referred-to-input (RTI) amplitude-threshold uncertainty in volts, and as a percent of the full-scale output signal V_{oFS} following amplification by A_{vdiff}. Error totals are derived from respective amplifier input parameter contributions defined in Equation (2-23), where barred quantities denote mean values and unbarred quantities refer to

Table 2-4. Amplifier error comparison (V_{diff} = 1 V, A_{vdiff} = 1, V_{oFS} = 1 V, dT = 10°C)

Parameter	OP-07	OPA-128	AD624	AD215	OPA-646	Comment
R_s	10 KΩ	10 MΩ	1 KΩ	50 Ω	75 Ω	Input group
V_{CM}	± 10 V	± 10 V	± 10 V	± 1000 V	± 10 V	
V_{OS}	$\overline{10}$ µV	$\overline{140}$ µV	$\overline{25}$ µV	$\overline{400}$ µV	$\overline{1000}$ µV	Offset group
$\dfrac{dV_{OS}}{dT} \cdot dT$	2 µV	50 µV	2.5 µV	20 µV	120 µV	
$I_{OS} \cdot R_s$	$\overline{3}$ µV	$\overline{0.3}$ µV	$\overline{10}$ µV	$\overline{15}$ µV	$\overline{30}$ µV	
$6.6\,Vn\sqrt{f_{hi}}$	51 µV	126 µV	26 µV	Negligible	1262 µV	Interference group
$\dfrac{V_{CM}}{CMRR}$	1000 µV	1000 µV	100 µV	10,000 µV	1000 µV	
$f(Av) \cdot \dfrac{V_{oFS}}{A_{vdiff}}$	$\overline{100}$ µV	$\overline{100}$ µV	$\overline{10}$ µV	$\overline{50}$ µV	$\overline{250}$ µV	Nonlinearity group
$\dfrac{dAV}{dT} \cdot dT \cdot \dfrac{V_{oFS}}{A_{vdiff}}$	500 µV	500 µV	50 µV	150 µV	500 µV	
$\varepsilon_{ampl\,\text{RTI}}$	$(\overline{113} + 1119)$ µV	$(\overline{240} + 1126)$ µV	$(\overline{45} + 115)$ µV	$(\overline{465} + 10{,}003)$ µV	$(\overline{1280} + 1690)$ µV	Σmean + 1σ RSS
$\varepsilon_{ampl\%FS}$	0.123% FS	0.136% FS	0.016% FS	1.046% FS	0.297% FS	$\times \dfrac{A_{vdiff}}{V_{oFS}} \times 100\%$

systematic and random values combined as the root sum square. Note that $A_{v_{diff}}$ normally is scaled for the V_{diff} input signal maximum in order to achieve a V_{oFS} of interest at the amplifier output. However, for the normalized examples of Table 2-4 each $A_{v_{diff}}$ is unity, requiring input V_{diff} values that equal the V_{oFS} value.

$$\varepsilon_{\text{ampl\%FS}} = \left\{\varepsilon_{\text{amplRTI volts}}\right\} \times \frac{A_{v_{diff}}}{V_{o_{FS}}} \cdot 100\%$$ (2-23)

$$= \left\{ \overline{V_{OS}} \; + \; \overline{I_{OS} \cdot R_s} \; + \; \overline{f(A_v) \cdot \frac{V_{o_{FS}}}{A_{v_{diff}}}} \right.$$

$$+ \left[\left(\frac{dV_{OS}}{dT} \cdot dT\right)^2 \; + \; \left(\frac{V_{CM}}{\text{CMRR}}\right)^2 \; + \; \left(6.6 \, \text{Vn}\sqrt{f_{hi}}\right)^2\right.$$

$$\left.\left. + \left(\frac{dAv}{dT} \cdot dT \cdot \frac{V_{o_{FS}}}{A_{v_{diff}}}\right)^2\right]^{1/2} \right\} \times \frac{A_{v_{diff}}}{V_{o_{FS}}} \cdot 100\%$$

Each amplifier is evaluated at identical $A_{v_{diff}}$, V_{oFS}, and temperature dT for consistency, but at expected R_s and V_{cm} values relevant to their typical application. All of the amplifiers are capable of accommodating off-ground and electromagnetically coupled V_{cm} input interference with an effectiveness determined by their respective $CMRR$, where the influence of amplifier $CMRR$ values in attenuating respective V_{cm} values is described. Mean offset voltages V_{os} are also untrimmed to reveal these possible differences. The OP-07 is assumed to be applied to an austere four-resistor subtractor circuit, resulting in its 10K R_s, whereas the OPA-128 low-input-bias-current amplifier interfaces a 10M R_s current sensor. The AD624 three-amplifier circuit offers the best performance and robustness overall, with its total error summation a tenth that of the other amplifiers.

The AD215 isolation amplifier 50 Ω R_s represents either the output of a preceding front-end instrumentation amplifier or low-level emf sensor. It is notable that the presence of a 1000 V V_{cm} input essentially accounts for the total error of this amplifier, which will be safely accommodated by the amplifier physical structure. Finally, with a 75 Ω coax R_s the wideband OPA-646 differs from other amplifiers in providing ten times the bandwidth at ten times the internal noise contribution. All of the amplifier error totals are commensurable owing to similar manufacturing technologies. Amplifier V_n rms internal noise voltage is converted to peak to peak with multiplication by 6.6, to account for its crest factor, dimensionally equivocating it to other amplifier input values in each error total.

PROBLEMS

2-1. The operational amplifier duplexer circuit shown provides directional coupling between a computer MODEM and a telephone line. Determine resistor values that maximize gains A_1 and A_2 while simultaneously minimizing A_3, and find the numerical magnitude of these gains.

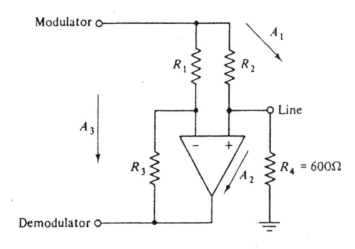

Figure 2-14.

2-2. Extend the amplifier error comparison of Table 2.4 with an evaluation of Analog Devices OP177E ultraprecision operational amplifier parameters obtained from the manufacturer's specifications. Utilize Equation (2-23) in this evaluation with an input R_s of 10 KΩ and V_{cm} equalling $\pm$ 10 V. Assume that amplifier gain drift with temperature is attributable to resistor temperature coefficients of 20 ppm per °C. Express error both as referred to input in μV and as a %FS. Consider an in-circuit *CMRR* numerical value of 100,000.

2-3. A designer has identical gain modules of gain K = 1000. Two amplifier topologies are to be investigated to provide a precise gain $G = y/r$ of 100.0. Investigate the given topologies and choose the one offering the highest performance minimizing output noise y/n for identical internal module noise n, following determination of the expression for G and B of each amplifier topology.

Topology I

$$\sqrt[3]{\frac{y}{r}} = \frac{K}{1+KB}$$

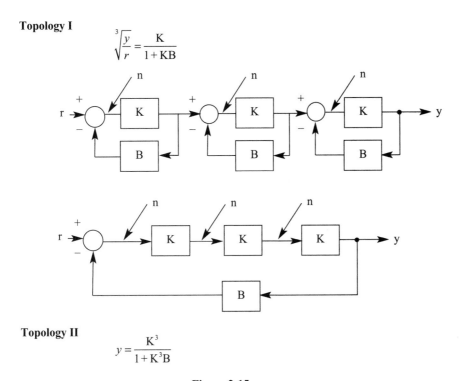

Topology II

$$y = \frac{K^3}{1+K^3B}$$

Figure 2-15.

BIBLIOGRAPHY

1. Bailey, D.C., "An Instrumentation Amplifier Is Not an Op Amp," *Electronic Products,* September 18, 1972.
2. Connelly, J.A., *Analog Integrated Circuits,* New York: Wiley-Interscience, 1975.
3. Embinder, J., *Application Considerations for Linear Integrated Circuits,* New York: Wiley-Interscience, 1970.
4. Fitchen, F.C., *Electronic Integrated Circuits and Systems,* New York: Van Nostrand Reinhold, 1970.

5. Garrett, P.H., *Analog I/O Design, Acquisition: Conversion: Recovery,* Reston, VA: Reston Publishing Co., 1981.

6. Graeme, J.G., *Applications of Operational Amplifiers: Third-Generation Techniques,* New York: McGraw-Hill, 1973.

7. Jaquay, J.W., "Designer's Guide to Instrumentation Amplifiers," *Electronic Design News,* May 5, 1972.

8. Kollataj, J.H., "Reject Common-Mode Noise," *Electronic Design,* April 26, 1973.

9. Lyerly, T.C., "Instrumentation Amplifier Conditions Computer Inputs," *Electronics,* November 6, 1972.

10. Netzer, Y., "The Design of Low-Noise Amplifiers," *Proceedings of IEEE,* June 1981.

11. Pettit, J.M and McWhorter, M.M, *Electronic Amplifier Circuits,* New York: McGraw-Hill, 1961.

12. Rutkowski, G.B., *Handbook of Integrated-Circuit Operational Amplifiers,* Englewood Cliffs, NJ: Prentice-Hall, 1975.

13. Tobey, G., Graeme, J., and Huelsman, L., *Operational Amplifiers: Design and Applications,* New York: McGraw-Hill, 1971.

CHAPTER 3

Filters for Measurement Signals

3-0   INTRODUCTION

Although the requirement for electric wave filters extends over an interval of a century, beginning with Marconi's radio experiments, the identification of stable and ideally terminated filter networks has occurred only during the past forty years. Filtering at the lower instrumentation frequencies has always been a problem with passive filters because the required L and C values are larger and inductor losses appreciable. The band-limiting of measurement signals in instrumentation applications imposes the additional concern of filter error additive to these measurement signals when accurate signal processing is required.

Consequently, this chapter provides a development of low-pass and band-pass filter characterizations appropriate for measurement signals, and develops filter error analyses for the more frequently required low-pass realizations. The excellent stability of active filter networks in the dc to 100-kHz instrumentation frequency range makes these circuits especially useful. Filter error analysis is accordingly developed to optimize the implementation of these filters for input signal conditioning, aliasing prevention, and output interpolation purposes associated with data conversion systems.

3-1   BANDLIMITING INSTRUMENTATION FILTERS

Low-pass filters are frequently required to band-limit measurement signals in instrumentation applications to achieve a frequency-selective function of interest.

Advanced Instrumentation and Computer I/O Design, Second Edition. By Patrick H. Garrett
Copyright © 2013 the Institute of Electrical and Electronics Engineers, Inc.

The application of an arbitrary signal set to a low-pass filter can result in a significant attenuation of higher frequency components, thereby defining a stop band whose boundary is influenced by the choice of filter cutoff frequency, with the unattenuated frequency components defining the filter passband. For instrumentation purposes, approximating the ideal low-pass filter amplitude $A(f)$ and phase $B(f)$ responses described by Figure 3-1 is beneficial in order to achieve signal band-limiting without alteration or the addition of errors to a passband signal of interest. In fact, preserving the accuracy of measurement signals is of sufficient importance that consideration of filter characterizations that correspond to well-behaved functions such as Butterworth and Bessel polynomials are especially useful. However, an ideal filter is physically unrealizable because practical filters are represented by ratios of polynomials that cannot possess the discontinuities required for sharply defined filter boundaries.

Figure 3-2 describes the Butterworth low-pass amplitude response $A(f)$ and Figure 3-3 its phase response $B(f)$, where n denotes the filter order or number of poles. Butterworth filters are characterized by a maximally flat amplitude response in the vicinity of dc, which extends towards its -3dB cutoff frequency f_c as n increases. This characteristic is defined by Equations (3-1) and (3-2) and Table 3-1. Butterworth attenuation is rapid beyond f_c as filter order increases, with a slightly nonlinear phase response that provides a good approximation to an ideal

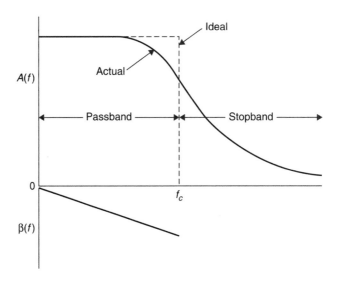

Figure 3-1. Ideal low-pass filter.

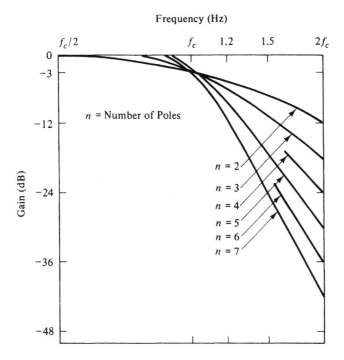

Figure 3-2. Butterworth low-pass amplitude.

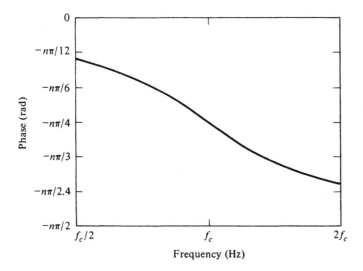

Figure 3-3. Butterworth low-pass phase.

Table 3-1. Butterworth polynomial coefficients

Poles, n	b_0	b_1	b_2	b_3	b_4	b_5
1	1.0					
2	1.0	1.414				
3	1.0	2.0	2.0			
4	1.0	2.613	3.414	2.613		
5	1.0	3.236	5.236	5.236	3.236	
6	1.0	3.864	7.464	9.141	7.464	3.864

low-pass filter. An analysis of the error attributable to this approximation is de-rived in Section 3-3. Figure 3-4 presents the Butterworth high-pass response.

$$B(s) = \left(j\frac{f}{f_c} \right)^n + b_{n-1}\left(j\frac{f}{f_c} \right)^{n-1} + \cdots + b_0 \tag{3-1}$$

$$A(f) = \frac{b_0}{\sqrt{B(s)B(-s)}} \tag{3-2}$$

$$= \frac{1}{\sqrt{1 + (f/f_c)^{2n}}}$$

Bessel filters are all-pole filters, like their Butterworth counterparts, with an amplitude response described by Equations (3-3) and (3-4) and Table 3-2. Bessel low-pass filters are characterized by a more linear phase delay extending to their cutoff frequency f_c and beyond as a function of filter order n shown in Figure 3-5. However, this linear-phase property applies only to low-pass filters. Unlike the flat passband of Butterworth low-pass filters, the Bessel passband has no value that does not exhibit amplitude attenuation with a Gaussian amplitude response described by Figure 3-6. It is also useful to compare the overshoot of Bessel and Butterworth fil-ters in Table 3-3, which reveals the Bessel to be much better behaved for band-lim-iting pulse-type instrumentation signals and where phase linearity is essential.

$$A(f) = \frac{b_0}{\sqrt{B(s)B(-s)}} \tag{3-3}$$

$$B(s) = \left(j\frac{f}{f_c} \right)^n + b_{n-1}\left(j\frac{f}{f_c} \right)^{n-1} + \cdots + b_0 \tag{3-4}$$

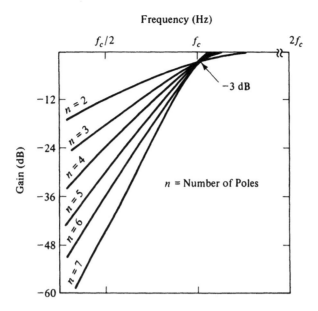

Figure 3-4. Butterworth high-pass amplitude.

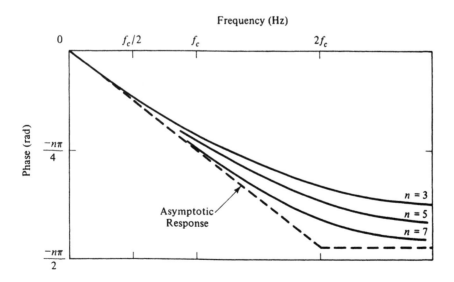

Figure 3-5. Bessel low-pass phase.

Table 3-2. Bessel polynomial coefficients

Poles, n	b_0	b_1	b_2	b_3	b_4	b_5
1	1					
2	3	3				
3	15	15	6			
4	105	105	45	10		
5	945	945	420	105	15	
6	10,395	10,395	4725	1260	210	21

3-2   ACTIVE FILTER DESIGN

In 1955, Sallen and Key of MIT published a description of 18 active filter networks for the realization of various filter approximations [11]. However, a rigorous sensitivity analysis by Geffe and others disclosed by 1967 that only four of the original networks exhibited low sensitivity to component drift. Of these, the unity-gain and multiple-feedback networks are of particular value for implementing low-pass and bandpass filters, respectively, to Q values of 10. Work by others resulted in the low-sensitivity biquad resonator, which can provide stable Q values to 200, and the sta-

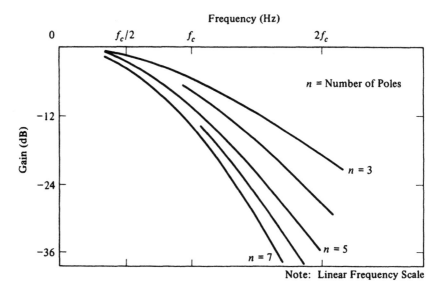

Figure 3-6. Bessel low-pass amplitude.

Table 3-3. Filter overshoot pulse response

n	Bessel (%FS)	Butterworth (%FS)
1	0	0
2	0.4	4
3	0.7	8
4	0.8	11

ble gyrator band-reject filter. These four networks are shown in Figure 3-7 with key sensitivity parameters. The sensitivity of a network can be determined, for example, when the change in its Q for a change in its passive-element values is evaluated. Equation (3-5) describes the change in the Q of a network by multiplying the thermal coefficient of the component of interest by its sensitivity coefficient. Normally, 50-to-100-ppm/°C components yield good performance.

$$S_Z^Q = \pm 1 \text{ passive network} \tag{3-5}$$

$$= (\pm 1)(50 \text{ ppm/°C})(100\%)$$

$$= \pm 0.005\% \ Q/°C$$

Unity-gain networks offer excellent performance for low-pass and highpass realizations and may be cascaded for higher-order filters. This is perhaps the most widely applied active filter circuit. Note that its sensitivity coefficients are less than unity for its passive components—the sensitivity of conventional passive networks—and that its resistor temperature coefficients are zero. However, it is sensitive to filter gain, indicating that designs that also obtain greater than unity gain with this filter network are suboptimum. The advantage of the multiple-feedback network is that a bandpass filter can be formed with a single operational amplifier, although the biquad network must be used for high-Q bandpass filters. However, the stability of the biquad at higher Q values depends upon the availability of adequate amplifier loop gain at the filter center frequency. Both bandpass networks can be stagger-tuned for a maximally flat passband response when required. The principle of operation of the gyrator is that a negative conductance G gyrates a capacitive current to an effective inductive current. Frequency stability is very good and a band-reject filter notch depth to about –40 dB is generally available. It should be appreciated that the principal capability of the active filter network is to synthesize a complex conjugate pole pair. This achievement, as described below, permits the realization of any mathematically definable low-pass approximation.

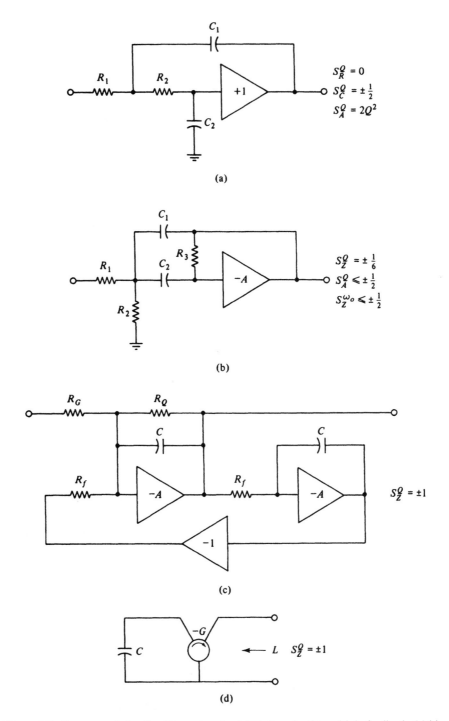

Figure 3-7. Recommended active filter networks. (a) Unity gain, (b) multiple feedback, (c) biquad, (d) gyrator.

Kirchoff's current law states that the sum of the currents into any node is zero. A nodal analysis of the unity-gain low-pass network yields Equations (3-6) through (3-9). It includes the assumption that current in C_2 is equal to current in R_2; the realization of this requires the use of a low-input-bias-current operational amplifier for accurate performance. The transfer function is obtained upon substituting for V_x in Equation (3-6) its independent expression obtained from Equation (3-7). Filter pole positions are defined by Equation (3-9). Figure 3-8 shows these nodal equations and the complex-plane pole positions mathematically described by Equation (3-9). This second-order network has two denominator roots (two poles) and is sometimes referred to as a resonator.

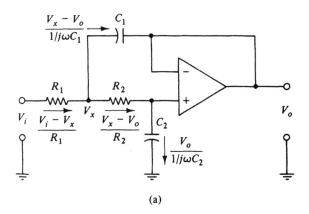

(a)

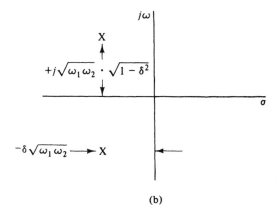

(b)

Figure 3-8. Unity-gain network nodal analysis.

$$\frac{V_i - V_x}{R_1} = \frac{V_x - V_o}{1/j\omega C_1} + \frac{V_x - V_o}{R_2} \tag{3-6}$$

$$\frac{V_x - V_o}{R_2} = \frac{V_o}{1/j\omega C_2} \tag{3-7}$$

After rearranging,

$$V_x = V_o \cdot \frac{R_2 + 1/j\omega C_2}{1/j\omega C_2}$$

$$\frac{V_o}{V_i} = \frac{1}{\omega^2 R_1 R_2 C_1 C_2 + \omega C_2 (R_1 + R_2) + 1} \tag{3-8}$$

$$\omega_1 = \frac{1}{R_1 C_1} \quad \text{and} \quad \omega_2 = \frac{1}{R_2 C_2}$$

$$\delta = \frac{C_2}{2}(R_1 + R_2)$$

$$S_{1,2} = \delta\sqrt{\omega_1 \omega_2} \pm j\sqrt{\omega_1 \omega_2} - \sqrt{1 - \delta^2} \tag{3-9}$$

A recent technique using MOS technology has made possible the realization of multipole unity-gain network active filters in total integrated-circuit form without the requirement for external components. Small-value MOS capacitors are utilized with MOS switches in a switched-capacitor circuit for simulating large-value resistors under control of a multiphase clock. With reference to Figure 3-9, the rate f_s at which the capacitor is toggled determines its charging to V and discharging to V'. Consequently, the average current flow I from V to V' defines an equivalent resistor R that would provide the same average current shown by the identity of Equation (3-10).

$$R = \frac{V - V'}{I} = 1/Cf_c \tag{3-10}$$

The switching rate f_s is normally much higher than the signal frequencies of interest so that the time sampling of the signal can be ignored in a simplified analysis.

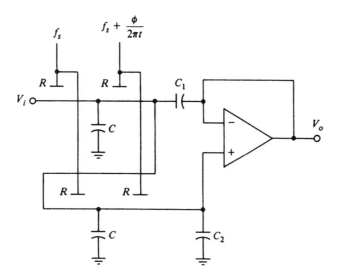

Figure 3-9. Switched capacitor unity-gain network.

Filter accuracy is primarily determined by the stability of the frequency f_s and the accuracy of implementation of the monolithic MOS capacitor ratios.

The most important parameter in the selection of operational amplifiers for active filter service is open-loop gain. The ratio of open-loop to closed-loop gain, or loop gain, must be 100 or greater for stable and well-behaved performance at the highest signal frequencies present. This is critical in the application of band-pass filters to ensure a realization that accurately follows the design calculations. Amplifier input and output impedances are normally sufficiently close to the ideal infinite input and zero output values to be inconsequential for impedances in active filter networks. Metal-film resistors having a temperature coefficient of 50 ppm/°C are recommended for active filter design.

Selection of capacitor type is the most difficult decision because of many interacting factors. For most applications, polystyrene capacitors are recommended because of their reliable –120 ppm/°C temperature coefficient and 0.05% capacitance retrace deviation with temperature cycling. Where capacitance values above 0.1 µF are required, however, polycarbonate capacitors are available in values to 1 µF with a ±50 ppm/°C temperature coefficient and 0.25% retrace. Mica capacitors are the most stable devices with ±50 ppm/°C temperature coefficient and 0.1% retrace, but practical capacitance availability is typically only 100 pF to 5000 pF. Mylar capacitors are available in values to 10µF with 0.3% retrace, but their temperature coefficient averages 400 ppm/°C.

The choice of resistor and capacitor tolerance determines the accuracy of the filter implementation, such as its cutoff frequency and passband flatness. Cost considerations normally dictate the choice of 1% tolerance resistors and 2 to 5% tolerance capacitors. However, it is usual practice to pair larger and smaller capacitor values to achieve required filter network values to within 1%, which results in filter parameters accurate to 1 or 2% with low temperature coefficient and retrace components. Filter response is typically displaced inversely to passive-component tolerance, such as lowering of cutoff frequency for component values on the high side of their tolerance band. For more critical realizations, such as high-Q bandpass filters, some provision for adjustment provides flexibility needed for an accurate implementation.

Table 3-4 provides the capacitor values in farads for unity-gain networks tabulated according to the number of filter poles. Higher-order filters are formed by a cascade of the second- and third-order networks shown in Figure 3-10, each of which is different. For example, a sixth-order filter will have six different capacitor values and not consist of a cascade of identical two-pole or three-pole networks. Figure 3-11 and 3-12 illustrate the design procedure with 1-kHz-cutoff two-pole Butterworth low-pass and highpass filters, including the frequency and impedance scaling steps. The three-pole filter design procedure is identical with

Table 3-4. Unity-gain network capacitor values in farads

Poles	Butterworth			Bessel		
	C1	C2	C3	C1	C2	C3
2	1.414	0.707		0.907	0.680	
3	3.546	1.392	0.202	1.423	0.988	0.254
4	1.082	0.924		0.735	0.675	
	2.613	0.383		1.012	0.390	
5	1.753	1.354	0.421	1.009	0.871	0.309
	3.235	0.309		1.041	0.310	
6	1.035	0.966		0.635	0.610	
	1.414	0.707		0.723	0.484	
	3.863	0.259		1.073	0.256	
7	1.531	1.336	0.488	0.853	0.779	0.303
	1.604	0.624		0.725	0.415	
	4.493	0.223		1.098	0.216	
8	1.091	0.981		0.567	0.554	
	1.202	0.831		0.609	0.486	
	1.800	0.556		0.726	0.359	
	5.125	0.195		1.116	0.186	

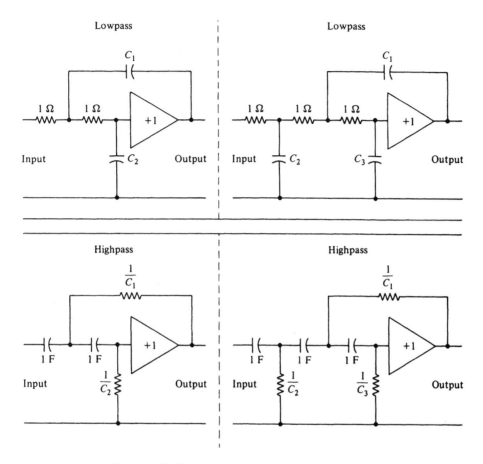

Figure 3-10. Two- and three-pole unity-gain networks.

observation of the appropriate network capacitor locations, but should be driven from a low driving-point impedance such as an operational amplifier. A design guide for unity-gain active filters is summarized in the following steps.

1. Select an appropriate filter approximation and number of poles required to provide the necessary response from the curves of Figures 3-2 through 3-6.
2. Choose the filter network appropriate for the required realization from Figure 3-10 and perform the necessary component frequency and impedance scaling.
3. Implement the filter components by selecting 1% standard-value resistors and then pairing a larger and smaller capacitor to realize each capacitor value to within 1%.

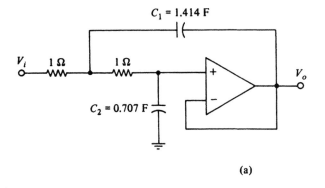

Component values from Table
3-4 are normalized to
1 rad/s with resistors taken
as 1 Ω and capcitors in farads.

(a)

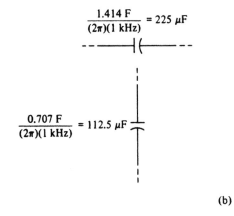

The filter is then
frequency-scaled by
dividing the capacitor
values from the table
by the cutoff frequency
in radians (2π X 1 kHz).

(b)

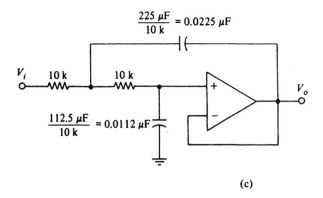

The filter is finally
impedance-scaled by
multiplying the resistor
values by a convenient
value (10 k) and dividing
the capacitor values by
the same value.

(c)

Figure 3-11. Buttworth unity-gain low-pass filter example.

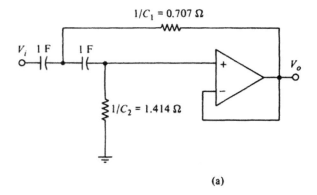

Component values from
Table 3-4 are normalized
to 1 rad/s with capacitors
taken as 1 F and resistors
the inverse capacitor values
from the table in ohms.

(a)

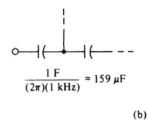

The filter is then frequency-
scaled by dividing the
capacitor values by the
cutoff frequency in radians
of value ($2\pi \times 1$ kHz).

(b)

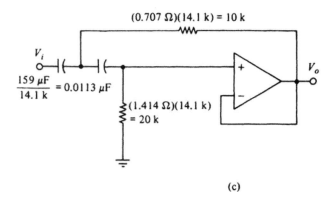

The filter is finally
impedance-scaled by
multiplying the resistor
values by a convenient
value (14.1 k) and
dividing the capacitor
values by the same
value.

(c)

Figure 3-12. Buttworth unity-gain high-pass filter example.

3-3   FILTER ERROR EVALUATION

Requirements for signal band-limiting in data acquisition and conversion systems include signal quality upgrading by signal conditioning circuits, aliasing prevention associated with sampled-data operations, and intersample error smoothing in output signal reconstruction. The accuracy, stability, and efficiency of low-pass active filter networks satisfy most of these requirements with the realization of filter characteristics appropriate for specific applications. However, when a filter is superimposed on a signal of interest, filter gain and phase deviations from the ideal result in a signal amplitude error that constitutes component error. It is therefore useful to evaluate filter parameters in order to select filter functions appropriate for signals of interest. It will be shown that applying this approach results in a minimum filter error added to the total system error budget. Since dc, sinusoidal, and harmonic signals are encountered in practice, analysis is performed for these signal types to identify optimum filter parameters for achieving minimum error.

$$\overline{\varepsilon_{\%FS}} \;=\; \frac{0.1}{f/f_c} \sum_{o}^{f/0.1f_c} [1.0 - A(f)] \cdot 100\% \qquad \text{dc and sinusoidal signals} \quad (3\text{-}11)$$

Both dc and sinusoidal signals exhibit a single spectral term. Filter gain error is thus the primary source of error because single-line spectra are unaffected by filter phase nonlinearities. Figure 3-13 describes the passband gain deviation, with reference to 0 Hz and expressed as average percent error of full scale, for three low-pass filters. The filter error attributable to gain deviation, $(1.0 - A(f))$ is shown to be minimum for the Butterworth characteristic, which is an expected result considering the passband flatness provided by Butterworth filters. Of significance is that small filter component errors can be achieved by restricting signal spectral occupancy to a fraction of the filter cutoff frequency.

Table 3-5 presents a tabulation of the example filters evaluated with dc and sinusoidal signals, defining mean amplitude errors for signal bandwidth occupancy to specified filter passband fractions of the cutoff frequency f_c. Equation (3-11) provides an approximate fit of filter parameters for RC, Bessel, and Butterworth filter characteristics. Most applications are better served by the three-pole Butterworth filter which offers a component error of $\overline{0.1\%FS}$ for signal passband occupancy to 40% of the filter cutoff, plus good stopband attenuation. Whereas it may appear inefficient not to utilize a filter passband up to its cutoff frequency,

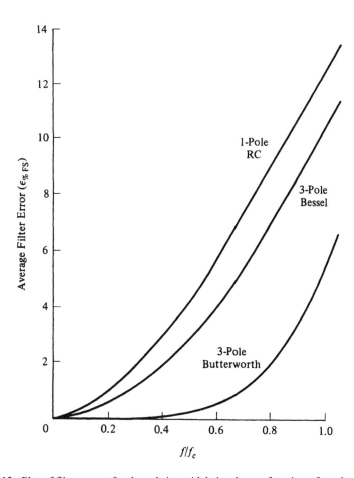

Figure 3-13. Plot of filter errors for dc and sinusoidal signals as a function of passband fraction.

the total bandwidth sacrificed is usually small. Higher filter orders may also be evaluated when greater stopband attenuation is of interest, with substitution of their amplitude response $A(f)$ in Equation (3-11).

The consequence of nonlinear phase delay with harmonic signals is described by Figure 3-14. The application of a harmonic signal just within the passband of a six-pole Butterworth filter provides the distorted output waveform shown. The variation in time delay between signal components at their specific frequencies results in a signal-time displacement and the amplitude alteration described. This time variation is apparent from evaluation of Equation (3-12),

Table 3-5. Filter amplitude errors for dc and sinusoidal signals

Signal bandwidth filter passband fractional occupancy	Amplitude response $A(f)$			Average filter error $\overline{\varepsilon}_{\%FS}$		
$\dfrac{f}{f_c}$	1-pole RC	3-pole Bessel	3-pole Butterworth	1-pole RC	3-pole Bessel	3-pole Butterworth
0.05	0.999	0.999	1.000	0.1	0.1	0
0.1	0.997	0.998	1.000	0.3	0.2	0
0.2	0.985	0.988	1.000	0.9	0.7	0
0.3	0.958	0.972	1.000	1.9	1.4	0
0.4	0.928	0.951	0.998	3.3	2.3	0.1
0.5	0.894	0.924	0.992	4.7	3.3	0.2
0.6	0.857	0.891	0.977	6.3	4.6	0.7
0.7	0.819	0.852	0.946	8.0	6.0	1.4
0.8	0.781	0.808	0.890	9.7	7.77	2.6
0.9	0.743	0.760	0.808	11.5	9.5	4.4
1.0	0.707	0.707	0.707	13.3	11.1	6.9

where linear phase provides a constant time delay. A comprehensive method for evaluating passband filter error for harmonic signals is reported by Brockman [2]. An error signal $\varepsilon(t)$ is derived as the difference between the output $y(t)$ of a filter of interest and a delayed input signal $x_0(t)$, expressed by Equations (3-13) through (3-15) and described in Figure 3-15. A volts-squared output error is then obtained from the Fourier transform of this error signal and the application of trigonometric identities, and expressed in terms of mean squared error (MSE) by Equation (3-16), with A_n and ϕ_n the filter magnitude and phase responses at n frequencies.

$$\text{Delay variation} = \frac{\phi_a}{2\pi f_a} - \frac{\phi_b}{2\pi f_b} \ \text{sec} \tag{3-12}$$

$$y(t) = \sum_{n=1}^{N} A_n \cos(\omega_n t - \phi_n) \tag{3-13}$$

$$x_0(t) = \sum_{n=1}^{N} \cos(\omega_n t - \omega_n t_0) \tag{3-14}$$

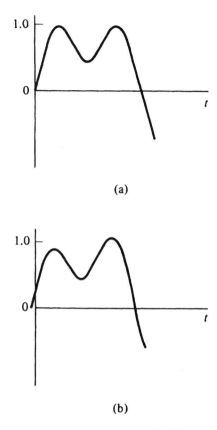

Figure 3-14. Filtered complex waveform phase nonlinearity: (a) sum of fundamental and third harmonic in 2:1 ratio; (b) sum of fundamental and third harmonic following six-pole low-pass Butterworth filter with signal occupancy to filter cutoff frequency.

$$\varepsilon(t) = y(t) - x_0(t) \tag{3-15}$$

$$= \sum_{n=1}^{N} [A_n \cos(\omega_n t - \phi_n) - \cos(\omega_n t - \omega_n t_0)]$$

$$\text{MSE} = \frac{1}{2} \sum_{n=1}^{N} [(A_n \cos\phi_n - \cos\omega_n t_0)^2 + (A_n \sin\phi_n - \sin\omega_n t_0)^2] \tag{3-16}$$

Computer simulation of first- through eighth-order Butterworth and Bessel low-pass filters were obtained with the structure of Figure 3-15. The signal delay

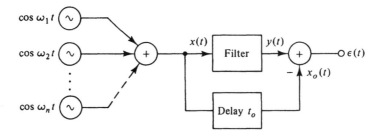

Figure 3-15. Filter harmonic signal error evaluation.

t_0 was varied in a search for the minimum true MSE by applying the Newton–Raphson method to the derivative of the MSE expression. This exercise was repeated for each filter with various passband spectral occupancies ranging from 10 to 100% of the cutoff frequency and $N = 10$ sinusoids per octave, represented as the simulated input signal $x(t)$. MSE is calculated by the substitution of each t_0 value in Equation (3-16), and expressed as average filter component error $\overline{\varepsilon_{\%FS}}$ by Equation (3-17) over the filter passband fraction specified for signal occupancy.

$$\overline{\varepsilon_{\%FS}} = \frac{\sqrt{\mathrm{MSE}}}{x(t)} \cdot 100\% \text{ harmonic signals} \qquad (3\text{-}17)$$

Table 3-6 provides a tabulation of these results describing an efficient filter-cutoff-to-signal-bandwidth ratio f_c/BW of 3, considering filter passband signal occupancy versus minimized component error. Signal spectral occupancy up to the filter cutoff frequency is also simulated for error reference purposes. The application of higher-order filters is primarily determined by the need for increased stop-band attenuation compared with the additional complexity and component precision required for their realization.

Low-pass band-limiting filters are frequently required by signal conditioning channels as illustrated in the following chapters, and especially for presampling antialiasing purposes plus output signal interpolation in sampled-data systems. Of interest is whether the intelligence represented by a signal is encoded in its amplitude values, phase relationships, or combined. Filter mean nonlinearity errors presented in Tables 3-5 and 3-6 describe amplitude deviations of filtered signals resulting from nonideal filter magnitude and phase characteristics. It is clear from these tabulations that Butterworth filters contribute nominal error to signal amplitudes when their passband cutoff frequency is derated to multiples of a sig-

Table 3-6. Filter amplitude errors for harmonic signals

Filter order (Poles)			Average Filter Error $\overline{\varepsilon_{\%FS}}$		
RC	Butterworth	Bessel	$f_c = 10$ BW	$f_c = 3$ BW	$f_c = $ BW
1			1.201%		
	2			1.093%	6.834%
		2		0.688%	6.179%
	3			0.115%	5.287%
		3		0.677%	6.045%
	4			0.119%	5.947%
		4		0.698%	6.075%
	5			0.134%	6.897%
		5		0.714%	6.118%
	6			0.153%	7.900%
		6		0.725%	6.151%
	7			0.172%	8.943%
		7		0.997%	6.378%
	8			0.195%	9.996%
		8		1.023%	6.299%

nal BW value. Three-pole and greater Butterworth low-pass filters of at least an f_c/BW of three therefore add negligible error for all signal types.

When signal phase accuracy is essential for phase-coherent applications, ranging from communications to audio systems, including matrixed home-theatre signals, then Bessel low-pass filters are advantageous. For example, if only signal phase is of interest, an examination of Figure 3-5 and Tables 3-5 and 3-6 reveal that derating a three-pole Bessel filter passband cutoff frequency to three times the signal BW achieves a very linear phase, but signal amplitude error approaches $\overline{1\%FS}$. However, error down to $\overline{0.1 - 0.2\%FS}$ in both amplitude and phase are provided for any signal type when this low-pass filter is derated on the order of ten times signal BW. At that operating point, Bessel filters behave as pure delay lines to the signal.

3-4 BANDPASS INSTRUMENTATION FILTERS

The bandpass filter passes a band of frequencies of bandwidth Δf centered at a frequency f_0 and attenuates all other frequencies. The quality factor Q of this filter is a measure of its selectivity and is defined by the ratio $f_0/\Delta f$. Also of interest is the geometric mean of the upper and lower –3-dB frequencies defining Δf, or $f_g =$

$f_u \cdot f_L$. Equations (3-18) and (3-19) present the amplitude function for a second-order bandpass filter in terms of these quantities, with amplitude response for various Q values plotted in Figure 3-16.

$$A(f) = \frac{2\pi f_0 / Q}{\sqrt{B(s)B(-s)}} \tag{3-18}$$

$$B(s) = (j2\pi f) + \frac{2\pi f_0}{Q} + \frac{(2\pi f_0)^2}{j2\pi f} \tag{3-19}$$

It may be appreciated from this figure that for all Q values the bandpass skirt attenuation rolloff relaxes to -12 dB/octave, one octave above and below f_0, which is expected for any second-order filter. (An octave is the interval between two fre-

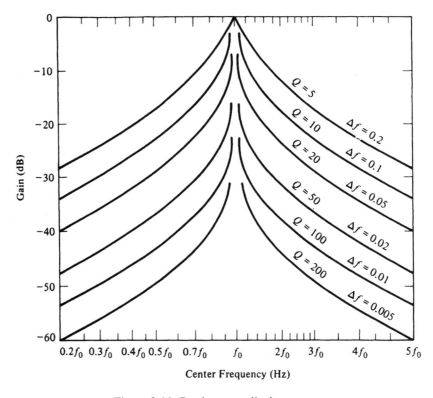

Figure 3-16. Band-pass amplitude response.

quencies, one twice the other.) Greater skirt attenuation can be obtained by cascading these single-tuned sections, thereby producing a higher-order filter. The phase response of a bandpass filter may be envisioned as that of a high-pass and low-pass filter in cascade. This phase has a slope whose change is monotonic and of value 0° at f_0, asymptotically reaching its maximum positive and maximum negative phase shift below and above f_0, respectively; total phase shift is a function of the filter order, $n \cdot 90°$.

The band-reject filter, also called a band-elimination or notch filter, passes all frequencies except those centered about f_c. Its amplitude function is described by Equations (3-20) to (3-22), and its amplitude response by Figure 3-17. Band-reject Q is determined by the ratio $f_c/\Delta f$, where bandwidth Δf is defined between the −3-dB passband cutoff frequencies. Band-reject-filter phase response follows the same phase characteristics described for the band-pass filter. For instrumentation service, the band-reject response can be obtained from the low-pass Butterworth coefficients of Table 3-1, and a maximally flat passband can be realized with paralleled Butterworth low-pass plus high-pass filters.

$$A(f) = \frac{1}{\sqrt{B(s)B(-s)}} \tag{3-20}$$

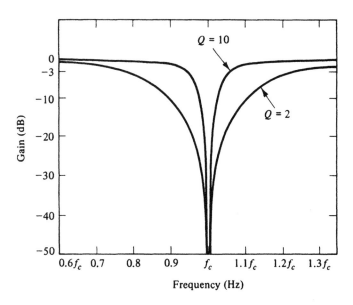

Figure 3-17. Band-reject amplitude response.

$$B(s) = C^n + b_{n-1}C^{n-1} + \ldots b_0 \tag{3-21}$$

$$C = \frac{\Delta f(j2\pi f)}{(j2\pi f)^2 + (2\pi f_c)^2} \tag{3-22}$$

A multiple-feedback band-pass filter (MFBF) is shown in Figure 3-18 with a center frequency of 1 Hz and a Q of 10. Equations (3-23) to (3-26) derive the component values for this filter. Normally, a standard capacitor value C is chosen in a range that results in reasonable resistor values with components selected to 1% tolerance. It should be noted that this circuit produces a signal inversion. When higher Q active band-pass filtering is required, the biquad network must be considered. Although its mechanization does require three operational amplifiers, the biquad provides the capability to independently set filter center frequency f_0, Q, and gain A_{f0} at the center frequency. A practical design approach is to fix the frequency-determining resistor R_{f0} shown in Figure 3-19 at a standard value, and then calculate the other component values as presented by the equations in Table 3-7 for representative instrumentation frequencies.

$$k = 2\pi f_0 C \tag{3-23}$$
$$= (6.28)(1 \text{ Hz})(1 \text{ }\mu\text{F})$$
$$= 6.28 \times 10^{-6} \text{ mho}$$

$$R_1 = \frac{Q}{k} \tag{3-24}$$
$$= \frac{10}{6.28 \times 10^{-6}}$$
$$= 1.6 \text{ M}$$

$$R_2 = \frac{1}{(2Q - 1/Q)k} \tag{3-25}$$
$$= \frac{1}{(20 - 0.1)(6.28 \times 10^{-6})}$$
$$= 8 \text{ }k$$

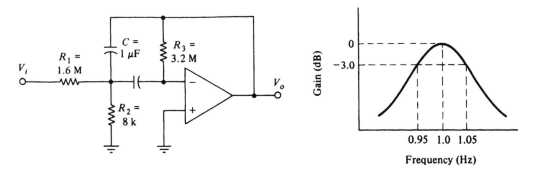

Figure 3-18. Multiple-feedback band-pass filter.

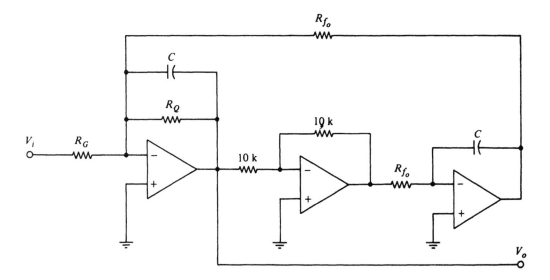

Figure 3-19. Biquad band-pass filter.

Table 3-7. Biquad component values for R_{f0} = 10 k

f_0	$C = \dfrac{1}{(2\pi f_o R_{f_o})}$	Q	$R_q = \dfrac{Q}{2\pi f_o C}$	A_{f_0}	$Rg = \dfrac{R_q}{A_{f_o}}$
10 Hz	1.6 μF	10	100 k	$Q/100$	1 M
100 Hz	0.16 μF	50	500 k	$Q/50$	500 k
1 kHz	0.016 μF	100	1 M	$Q/10$	100 k
10 kHz	0.0016 μF	200	2 M	Q	10 k

$$R_3 = \frac{2Q}{k} \qquad (3\text{-}26)$$

$$= \frac{20}{6.28 \times 10^{-6}}$$

$$= 3.2 \text{ M}$$

Most instrumentation systems involve amplitude measurements of transducer outputs, and it is normally of interest to maintain amplitude flatness in the signal passband. In the case of band-pass filtering using the previous single-tuned networks, the amplitude response rolls off immediately on both sides of the center frequency. Band-pass signals having an extended spectral occupancy, however, should be filtered by a flat-passband band-pass filter. A stagger-tuning scheme for multiple-cascaded, single-tuned band-pass filters can produce a flat passband with the additional benefit of increased skirt selectivity. Table 3-8 presents stagger-tuning parameters for a maximally flat passband in terms of the number of single-tuned networks employed, their individual center frequencies f_r and –3-dB bandwidths Δfr, and the overall band-pass filter center frequency f_0 and –3-dB bandwidth Δf. Passband flatness and skirt selectivity both improve, of course, as the number of cascaded single-tuned networks increases and the overall Δf decreases.

Consider, for example, a band-pass filter requirement centered at an f_0 of 1 kHz with a maximally flat Δf bandwidth of 200 Hz. This $Q = 5$ filter is also to achieve –35-dB attenuation ±1 octave on both sides of the center frequency f_0. Two cascaded and stagger-tuned MFBF networks are able to meet these specifications, requir-

Table 3-8. Stagger-tuning parameters

Single-tuned filters	Δf_r	f_r
2	0.71 Δf	$f_0 + 0.35\ \Delta f$
	0.71 Δf	$f_0 - 0.35\ \Delta f$
3	0.5 Δf	$f_0 + 0.43\ \Delta f$
	0.5 Δf	$f_0 - 0.43\ \Delta f$
	1.0 Δf	f_0
4	0.35 Δf	$f_0 + 0.46\ \Delta f$
	0.38 Δf	$f_0 - 0.46\ \Delta f$
	0.93 Δf	$f_0 + 0.19\ \Delta f$
	0.93 Δf	$f_0 - 0.19\ \Delta f$

ing only two operational amplifiers for their implementation. The individual MFBF networks are designed according to the example associated with Figure 3-18, but employing the tuning parameters obtained from Table 3-8. The filter circuit is shown by Figure 3-20 with 0.1 μF capacitors. In the event that final minor tuning adjustments are required, each network center frequency is determined by R_2, Q by R_3, and gain by R_1. A penalty of the stagger-tuned method is a gain loss that results from the algebraic addition of the skirts of each network. However, this loss may be calculated and compensated for on a per-network basis as shown in the band-pass filter example calculations that follow [Equations (3-27) through (3-30)]. The overall filter response achieved is described by Figure 3-21 and

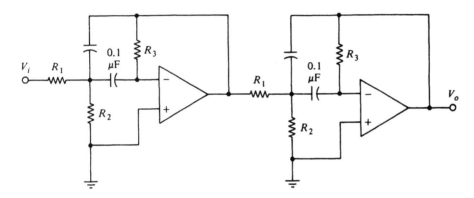

Figure 3-20. Stagger-tuned multiple-feedback band-pass filter.

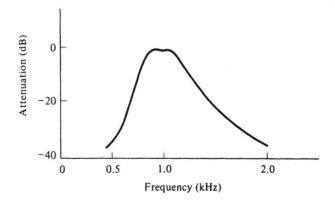

Figure 3-21. Stagger-tuned $Q = 5$ band-pass response.

has a +0.3-dB amplitude response at ±Hz (70% bandwidth) on either side of the 0-dB f_0.

$$\text{Gain loss}_r = \frac{A_r}{\sqrt{A_r^2 + B_r^2}} \tag{3-27}$$

$$A_r = \frac{(2\pi f_r)(2\pi f_g)}{Q_r} \tag{3-28}$$

$$B_r = (2\pi f_r)^2 - (2\pi f_g)^2 \tag{3-29}$$

$$f_g = \sqrt{f_u \cdot f_L} \tag{3-30}$$

$$= \sqrt{(1{,}100 \text{ Hz})(900 \text{ Hz})}$$

$$= 995 \text{ Hz}$$

First Section		**Second Section**	
$\Delta f_{r_1} = 0.71\,\Delta f$ $\quad= (0.71)(200 \text{ Hz})$ $\quad= 141 \text{ Hz}$	Table (3-8)	$\Delta f_{r_2} = 0.71\,\Delta f$ $\quad= (0.71)(200 \text{ Hz})$ $\quad= 141 \text{ Hz}$	Table (3-8)
$f_{r_1} = f_o + 0.35\,\Delta f$ $\quad= 1 \text{ kHz} + (0.35)(200 \text{ Hz})$ $\quad= 1.07 \text{ kHz}$	Table (3-8)	$f_{r_2} = f_o - 0.35\,\Delta f$ $\quad= 1 \text{ kHz} - (0.35)(200 \text{ Hz})$ $\quad= 930 \text{ kHz}$	Table (3-8)
$Q_1 = \dfrac{f_{r_1}}{\Delta f_{r_1}}$ $\quad= \dfrac{1.07 \text{ kHz}}{141 \text{ Hz}}$ $\quad= 7.6$	Figure 3-16	$Q_2 = \dfrac{f_{r_2}}{\Delta f_{r_2}}$ $\quad= \dfrac{930 \text{ kHz}}{141 \text{ Hz}}$ $\quad= 6.6$	Figure 3-16
$k_1 = 2\pi f_{r_1} C$ $\quad= (2\pi)(1.07 \text{ kHz})(0.1 \text{ μF})$ $\quad= 6.72 \times 10^{-4} \text{ mho}$	(3-23)	$k_2 = 2\pi f_{r_2} C$ $\quad= (2\pi)(930 \text{ Hz})(0.1 \text{ μF})$ $\quad= 5.85 \times 10^{-4} \text{ mho}$	(3-23)

$$A_1 = \frac{(2\pi f_{r_1})(2\pi f_g)}{Q_1} \qquad (3\text{-}28)$$

$$= \frac{(2\pi \cdot 1.07 \text{ kHz})(2\pi \cdot 995 \text{ Hz})}{7.6}$$

$$= 5.53 \times 10^6$$

$$B_1 = (2\pi f_{r_1})^2 - (2\pi f_g)^2 \qquad (3\text{-}29)$$
$$= (2\pi \cdot 1.07 \text{ kHz})^2 - (2\pi \cdot 995 \text{ Hz})^2$$
$$= 6.15 \times 10^6$$

$$\begin{matrix} \text{Gain} \\ \text{loss}_1 \end{matrix} = \frac{A_1}{\sqrt{A_1^2 + B_1^2}} \qquad (3\text{-}27)$$

$$= \frac{5.53 \times 10^6}{\sqrt{(5.53 \times 10^6)^2 + (6.15 \times 10^6)^2}}$$

$$= 0.669$$

$$R_1 = \frac{Q_1 \cdot \text{gain loss}_1}{K_1} \qquad (3\text{-}24)$$

$$= \frac{(7.6)(0.669)}{6.72 \times 10^{-4} \text{ mho}}$$

$$= 7.55 \text{ k}$$

$$R_2 = \frac{1}{\left(2Q_1 - \dfrac{1}{Q_1 \cdot \text{gain loss}_1}\right)K_1} \qquad (3\text{-}25)$$

$$= \frac{1}{\left(15.2 - \dfrac{1}{(7.6)(0.669)}\right)(6.72 \times 10^{-4}) \text{mho}}$$

$$= 99 \ \Omega$$

$$R_3 = \frac{2Q_1}{K_1} \qquad (3\text{-}26)$$

$$= \frac{15.2}{6.72 \times 10^{-4} \text{ mho}}$$

$$= 22.6 \text{ k}$$

$$A_2 = \frac{(2\pi f_{r_2})(2\pi f_g)}{Q_2} \qquad (3\text{-}28)$$

$$= \frac{(2\pi \cdot 930 \text{ Hz})(2\pi \cdot 995 \text{ Hz})}{6.6}$$

$$= 5.53 \times 10^6$$

$$B_2 = (2\pi f_{r_2})^2 - (2\pi f_g)^2 \qquad (3\text{-}29)$$
$$= (2\pi \cdot 930 \text{ Hz})^2 - (2\pi \cdot 995 \text{ Hz})^2$$
$$= -4.9 \times 10^6$$

$$\begin{matrix} \text{Gain} \\ \text{loss}_2 \end{matrix} = \frac{A_2}{\sqrt{A_2^2 + B_2^2}} \qquad (3\text{-}27)$$

$$= \frac{5.53 \times 10^6}{\sqrt{(5.53 \times 10^6)^2 + (-4.9 \times 10^6)^2}}$$

$$= 0.75$$

$$R_1 = \frac{Q_2 \cdot \text{gain loss}_2}{K_2} \qquad (3\text{-}24)$$

$$= \frac{(6.6)(0.75)}{5.85 \times 10^{-4} \text{ mho}}$$

$$= 8.47 \text{ k}$$

$$R_2 = \frac{1}{\left(2Q_2 - \dfrac{1}{Q_2 \cdot \text{gain loss}_2}\right)K_2} \qquad (3\text{-}25)$$

$$= \frac{1}{\left(13.2 - \dfrac{1}{(6.6)(0.75)}\right)(5.85 \times 10^{-4}) \text{mho}}$$

$$= 132 \ \Omega$$

$$R_3 = \frac{2Q_2}{K_2} \qquad (3\text{-}26)$$

$$= \frac{13.2}{5.85 \times 10^{-4} \text{ mho}}$$

$$= 22.6 \text{ k}$$

PROBLEMS

3-1. Design a seven-pole Butterworth low-pass filter, employing unity-gain networks that are frequency scaled for a 1-Hz cutoff frequency and using 1 M ohm resistors for impedance scaling. Sketch the final circuit including all component values, and provide a Bode plot of amplitude attenuation in dB versus frequency.

3-2. Implement a band-reject filter employing three-pole, unity-gain Butterworth low-pass and high-pass networks connected in the topology shown below, and terminated with an output follower amplifier. Use 100 KΩ resistors to provide filter impedance scaling with a 60 Hz notch frequency. Show the final circuit and component values.

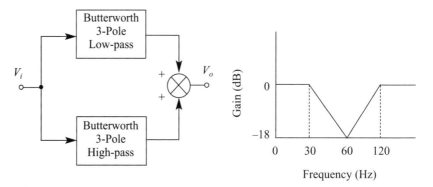

Figure 3-22.

3-3. A telemetry channel centered at 68 KHz is to be band-pass filtered over a maximally flat 16 KHz bandwith Δf, between f_L and f_u frequencies of 60 KHz and 76 KHz, achieving -30 dB attenuation at 53 KHz and 93 KHz. Implement three stagger-tuned MFBF networks employing Table 3-8 to achieve this response employing 0.001 μF capacitors. Sketch the filter circuits, including component values, show all calculations, and a composite plot of the overall response aided by Figure 3-16 applied to each network.

3-4. Design a Bessel band-pass filter with 800 Hz for 2 f_c and 1.5 kHz for $f_c/2$ -3dB frequencies using cascaded four-pole high-pass and four-pole low-pass filters, respectively. Show filter calculations using 10 KΩ impedance scaling resistors throughout, the final circuit, and a Bode plot of amplitude attenuation in dB versus frequency two octaves below 800 Hz and above 1.5 kHz, aided by Figure 3.6.

BIBLIOGRAPHY

1. Allen, P.E. and Huelsman, L.P., *Theory and Design of Active Filters* (New York, Wiley, 1975).
2. Brockman, J.P., "Interpolation Error in Sampled Data Systems," Electrical Engineering Department, University of Cincinnati, May 1985.
3. Craig, J.W., *Design of Lossy Filters* (Cambridge, MA: MIT Press, 1970).
4. Daniels, R.W., *Approximation Methods for Electronic Filter Design* (New York, McGraw-Hill, 1974).
5. Geffe, P.R., "Toward High Stability in Active Filters," *IEEE Spectrum,* Vol. 7, May 1970.
6. Johnson, D.E., *Introduction to Filter Theory* (Englewood Cliffs, NJ: Prentice-Hall, 1976).
7. Johnson, D.E and Hilburn, J.L., *Rapid Practical Designs of Active Filters* (New York, John Wiley, 1975).
8. Laube, S., "Comparative Analysis of Total Average Filter Component Error," Senior Design Project, Electrical Engineering Technology, University of Cincinnati, 1983.
9. Mitra, C., *Analysis and Synthesis of Linear Active Networks* (New York, Wiley, 1969).
10. Rhodes, J.D., *Theory of Electrical Filters* (New York: Wiley, 1976).
11. Sallen, R.P. and Key, E.L., "A Practical Method of Designing RC Active Filters," *IRE Transactions on Circuit Theory,* Vol. CT-2, March 1955.
12. Thomas, L.C., "The Biquad, Part 1—Some Practical Design Considerations," *IEEE Circuit Theory Transactions,* Vol. CT-18, May 1971.
13. Zeines, B., *Introduction to Network Analysis* (Englewood Cliffs: NJ: Prentice-Hall, 1967).

CHAPTER 4

Signal Conditioning Design and Instrumentation Errors

4-0   INTRODUCTION

Economic considerations are imposing increased accountability on the design of analog I/O systems to provide performance at the required accuracy for computer-integrated measurement and control instrumentation without the costs of overdesign. Within that context, this chapter describes the development of signal acquisition and conditioning systems, and derives a unified method for representing and upgrading the quality of instrumentation signals between sensors and data-conversion systems. Low-level signal conditioning is comprehensively developed for both coherent and random interference conditions employing sensor–amplifier–filter structures for signal quality improvement presented in terms of detailed device and system error budgets. Examples for dc, sinusoidal, and harmonic signals are provided, including grounding, shielding, and circuit noise considerations. A final section explores the additional signal quality improvement available by averaging redundant signal conditioning channels, including reliability enhancement. A distinction is made between signal conditioning, which is primarily concerned with operations for improving signal quality, and signal processing operations that assume signal quality at the level of interest.

4-1   LOW-LEVEL SIGNAL ACQUISITION

The designer of high-performance instrumentation systems has the responsibility of defining criteria for determining preferred options from among available alternatives. Figure 4-1 illustrates a cause-and-effect outline of methods that are developed in this chapter, whose application aids the realization of effective signal conditioning designs. In this fishbone chart, grouped system and device options are outlined for contributing to the goal of minimum total instrumentation error. Sensor choices appropriate for measurands of interest were introduced in Chapter 1 including linearization and calibration. Application-specific amplifier and filter choices for signal conditioning were defined, respectively, in Chapters 2 and 3. In this section, input circuit noise, impedance, and grounding effects are described for signal conditioning optimization. The following section derives models that combine device and system quantities in the evaluation and improvement of signal quality, expressed as total error, including the influence of random and coherent interference. The remaining chapter sections provide detailed examples of these signal conditioning design methods.

External interference entering low-level instrumentation circuits frequently is substantial and techniques for its attenuation are essential. Noise coupled to signal cables and power buses has as its cause electric and magnetic field sources. For example, signal cables will couple 1 mV of interference per kilowatt of 60-Hz load for each linear foot of cable run of 1-ft spacing from adjacent power cables. Most interference results from near-field sources, primarily electric fields, for which an effective attenuation mechanism is reflection by nonmagnetic materials such as copper or aluminum shielding. Both foil and braided shielded twinax signal cables offer attenuation on the order of -90 voltage dB to 60-Hz interference.

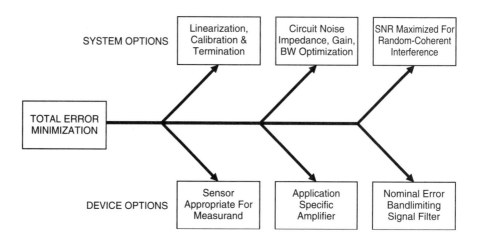

Figure 4-1. Signal conditioning design influences.

For magnetic fields, absorption is the effective attenuation mechanism requiring steel or mu-metal shielding. Magnetic fields are more difficult to shield than electric fields; shielding effectiveness for a specific thickness diminishes with decreasing frequency. For example, at 60 Hz steel provides interference attenuation on the order of −30 voltage dB per 100 mils of thickness. Applications requiring magnetic shielding are usually implemented by the installation of signal cables in steel conduit of the necessary wall thickness. Additional magnetic-field attenuation is furnished by periodic transposition of twisted-pair signal cable, provided no signal returns are on the shield; low-capacitance cabling is preferable. Mutual coupling between computer data-acquisition system elements, for example from finite ground impedances shared among different circuits, also can be significant, with noise amplitudes equivalent to 50 mV at signal inputs. Such coupling is minimized by separating analog and digital circuit grounds into separate returns to a common low-impedance chassis star-point termination, as illustrated in Figure 4-3.

The goal of shield ground placement in all cases is to provide a barrier between signal cables and external interference from sensors to their amplifier inputs. Signal cable shields also are grounded at a single point, below 1 MHz signal bandwidths, and ideally at the source of greatest interference, where provision of the lowest impedance ground is more beneficial. One instance in which a shield is not grounded is when driven by an amplifier guard. Guarding neutralizes cable-to-shield capacitance imbalance by driving the shield with common-mode interference appearing on the signal leads; this is also known as active shielding.

The components of total input noise may be divided into external contributions associated with the sensor circuit, and internal amplifier noise sources referred to its input. We shall consider the combination of these noise components in the context of band-limited sensor-amplifier signal acquisition circuits. Phenomena associated with the measurement of a quantity frequently involve energy–matter interactions that result in additive noise. Thermal noise V_t is present in all elements containing resistance above absolute zero temperature. Equation (4-1) defines thermal noise voltage proportional to the square root of the product of the source resistance and its temperature. This equation is also known as the Johnson formula which is typically evaluated at room temperature, or 293°K, and represented as a voltage generator in series with a noiseless source resistance.

$$V_t = \sqrt{4kTR_s} \; V_{rms} / \sqrt{\text{Hz}} \qquad (4\text{-}1)$$

where

k = Boltzmann's constant (1.38×10^{-23} J/°K)

T = absolute temperature (°K)

R_s = source resistance (Ω)

Thermal noise is not influenced by current flow through an associated resistance. However, a dc current flow in a sensor loop may encounter a barrier at any contact or junction connection that can result in contact noise owing to fluctuating conductivity effects. This noise component has a unique characteristic that varies as the reciprocal of signal frequency $1/f$, but is directly proportional to the value of dc current. The behavior of this fluctuation with respect to a sensor-loop source resistance is to produce a contact noise voltage whose magnitude may be estimated at a signal frequency of interest by the empirical relationship of Equation (4-2). An important conclusion is that dc current flow should be minimized in the excitation of sensor circuits, especially for low signal frequencies.

$$V_c = (0.57 \times 10^{-9})\, R_s \sqrt{\frac{I_{dc}}{f}}\; V_{rms}/\sqrt{\text{Hz}} \qquad (4\text{-}2)$$

where

I_{dc} = average dc current (A)

f = signal frequency (Hz)

R_s = source resistance (Ω)

Instrumentation amplifier manufacturers use the method of equivalent noise-voltage and noise-current sources applied to one input to represent internal noise sources referred to amplifier input, as illustrated in Figure 4-2. The short-circuit rms input noise voltage V_n is the random disturbance that would appear at the input of a noiseless amplifier, where its increase below 100 Hz is due to internal amplifier $1/f$ contact noise sources. The open-circuit rms input noise current I_n similarly arises from internal amplifier noise sources and usually may be disregarded in sensor-amplifier circuits because its small magnitude typically results in a negligible input disturbance, except when large source resistances are present. Since all of these input noise contributions are essentially from uncorrelated sources, they are combined as the root-sum-square by Equation (4-3). Wide bandwidths and large source resistances, therefore, should be avoided in sensor-amplifier signal acquisition circuits in the interest of noise minimization. Further, additional noise sources encountered in an instrumentation channel following the input gain stage are of diminished consequence because of noise amplification provided by the input stage.

$$V_{N_{pp}} = \left[6.6 \left(V_t^2 + V_c^2 + V_n^2 \right) \left(f_{hi} \right) \right]^{1/2} \tag{4-3}$$

4-2   SIGNAL QUALITY IN RANDOM AND COHERENT INTERFERENCE

The acquisition of a low-level analog signal that represents some measurand in the presence of appreciable interference is a frequent requirement. Of concern is achieving a signal amplitude measurement A or phase angle ϕ at the accuracy of interest through upgrading the quality of the signal by means of appropriate signal conditioning circuits. Closed-form expressions are available, notably by Raemer [9], for determining the error of a signal corrupted by random Gaussian noise or coherent sinusoidal interference. These are expressed in terms of signal-to-noise ratios (SNR) by Equations (4-4) through (4-9). SNR is a dimensionless ratio of watts of signal to watts of noise, and frequently is expressed as rms signal-to-interference amplitude squared. These equations are exact for sinusoidal signals, which are typical for excitations encountered with instrumentation sources.

$$P(\Delta A; A) = \mathrm{erf} \left(\frac{1}{2} \frac{\Delta A}{A} \sqrt{\mathrm{SNR}} \right) \text{ probability} \tag{4-4}$$

$$0.68 = \mathrm{erf} \left(\frac{1}{2} \frac{\varepsilon_{\%FS}}{100\%} \sqrt{\mathrm{SNR}} \right)$$

$$\varepsilon_{\text{random amplitude}} = \frac{\sqrt{2}}{\sqrt{\mathrm{SNR}}} \frac{100\%}{} \text{ of full scale } (1\sigma) \tag{4-5}$$

$$P(\Delta\varphi; \varphi) = \mathrm{erf} \left(\frac{1}{2} \frac{\Delta\varphi}{\varphi} \sqrt{\mathrm{SNR}} \right) \text{ probability} \tag{4-6}$$

$$0.68 = \mathrm{erf} \left(\frac{1}{2} \frac{\varepsilon_\varphi}{57.3^\circ / \mathrm{rad}} \sqrt{\mathrm{SNR}} \right)$$

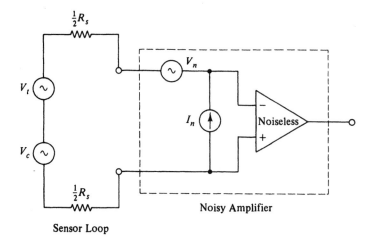

V_t = Thermal Noise
V_c = Contact Noise
I_n = Device Noise Current
V_n = Device Noise Voltage

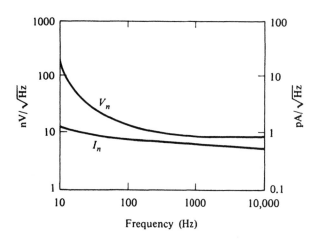

Figure 4-2. Sensor–amplifier noise sources.

$$\varepsilon_{\text{random phase}} = \frac{1}{2} \frac{\sqrt{2} \ 100}{\sqrt{\text{SNR}}} \text{ degrees } (1\sigma) \tag{4-7}$$

$$\varepsilon_{\text{coh amplitude}} = \frac{\Delta A}{A} \cdot 100\% \tag{4-8}$$

$$= \sqrt{\frac{V_{coh}^2}{V_{FS}^2}} \cdot 100\%$$

$$= \frac{100\%}{\sqrt{\text{SNR}}} \text{ of full scale}$$

$$\varepsilon_{\text{coh phase}} = \frac{100}{2 \ \sqrt{\text{SNR}}} \text{ degrees} \tag{4-9}$$

The probability that a signal corrupted by random Gaussian noise is within a specified Δ region centered on its true amplitude A or phase ϕ values is defined by Equations (4-4) and (4-6). Table 4-1 presents a tabulation from substitution into these equations for amplitude and phase errors at a 68% (1σ) confidence in their measurement for specific SNR values. One sigma is an acceptable confi-

Table 4-1. SNR versus amplitude and phase errors

SNR	Amplitude error, random $\varepsilon_{\%\text{FS}}$	Phase error, random $\varepsilon_{\phi\text{deg}}$	Amplitude error, coherent $\varepsilon_{\%\text{FS}}$
10^1	44.0	22.3	31.1
10^2	14.0	7.07	9.9
10^3	4.4	2.23	3.1
10^4	1.4	0.707	0.990
10^5	0.44	0.223	0.311
10^6	0.14	0.070	0.099
10^7	0.044	0.022	0.0311
10^8	0.014	0.007	0.0099
10^9	0.0044	0.002	0.0031
10^{10}	0.0014	0.0007	0.00099
10^{11}	0.00044	0.0002	0.00031
10^{12}	0.00014	0.00007	0.00009

dence level for many applications. For 95% (2σ) confidence, the error values are doubled for the same SNR. These amplitude and phase errors are closely approximated by the simplifications of Equations (4-5) and (4-7), and are more readily evaluated than by Equations (4-4) and (4-6). For coherent interference, Equations (4-8) and (4-9) approximate amplitude and phase errors, where ΔA is directly proportional to V_{coh}, as the true value of A is to V_{FS}. Errors due to coherent interference are seen to be less than those due to random interference by the $\sqrt{2}$ for identical SNR values. Further, the accuracy of all of these error expressions requires minimum SNR values of one or greater. This is usually readily achieved in practice by the associated signal conditioning circuits illustrated in the examples that follow. Ideal matched-filter signal conditioning makes use of both amplitude and phase information in upgrading signal quality, and is implied in these SNR relationships for amplitude and phase error in the case of random interference.

For practical applications, the SNR requirements ascribed to amplitude and phase error must be mathematically related to conventional amplifier and linear filter signal-conditioning circuits. Figure 4-3 describes the basic signal-conditioning structure, including a preconditioning amplifier and postconditioning filter and their bandwidths. Earlier work by Fano [4] showed that under high-input SNR conditions, linear filtering approaches matched filtering in its efficiency. Later work by Budai [2] developed a relationship for this efficiency expressed by the characteristic curve of Figure 4-4. This curve and its k parameter appears most reliable for filter numerical input SNR values between about 10 and 100, with an efficiency k of 0.9 for SNR values of 200 and greater.

Equations (4-10) through (4-13) describe the relationships with which the improvement in signal quality may be determined. Both rms and dc voltage values are interchangeable in Equation (4-10). The R_{cm} and R_{diff} impedances of the amplifier input termination account for the V^2/R transducer–gain relationship of the input SNR in Equation (4-11). CMRR is squared in this equation in order to convert its ratio of differential to common-mode voltage gains to a dimensionally correct power ratio. Equation (4-12) represents the narrower bandwidth processing–gain relationship for the ratio of amplifier f_{hi} to filter f_c producted with the filter efficiency k, for improving signal quality above that provided by the amplifier CMRR with random interference. Most of the improvement is provided by the amplifier CMRR, owing to its squared factor, but random noise higher-frequency components are also effectively attenuated by linear filtering.

$$\text{Input SNR} = \left(\frac{V_{\mathrm{diff}}}{V_{\mathrm{cm}}}\right)^2 \text{ dc or rms} \qquad (4\text{-}10)$$

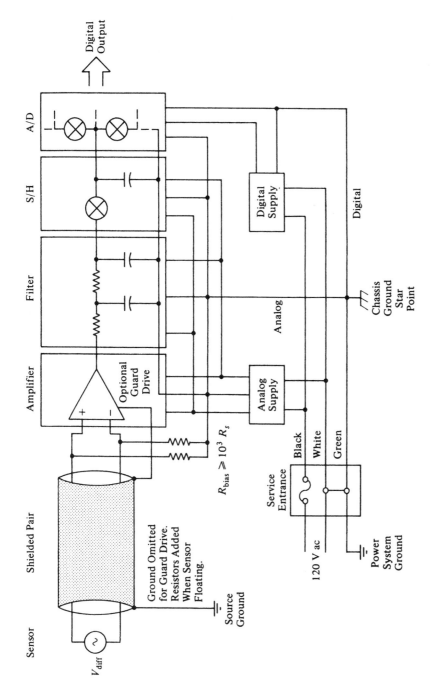

Figure 4-3. Signal acquisition system interfaces.

93

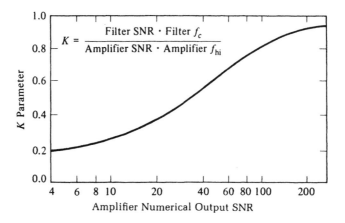

Figure 4-4. Linear filter efficiency K versus SNR.

$$\text{Amplifier SNR} = \text{input SNR} \cdot \frac{R_{cm}}{R_{diff}} \cdot \text{CMRR}^2 \qquad (4\text{-}11)$$

$$\text{Filter SNR}_{random} = \text{amplifier SNR} \cdot k \cdot \frac{f_{hi}}{f_c} \qquad (4\text{-}12)$$

$$\text{Filter SNR}_{coherent} = \text{amplifier SNR} \cdot \left[1 + \left(\frac{f_{coh}}{f_c} \right)^{2n} \right] \qquad (4\text{-}13)$$

For coherent interference conditions, signal-quality improvement is a function of achievable filter attenuation at the interfering frequencies. This is expressed by Equation (4-13) for one-pole RC to n-pole Butterworth low-pass filters. Note that filter cutoff frequency is determined from the considerations of Tables 3-5 and 3-6 with regard to minimizing the filter-component error contribution. Finally, the various signal-conditioning device errors and output signal qualities must be appropriately combined in order to determine total channel error. Sensor nonlinearity, amplifier, and filter errors are combined with the root-sum-square of signal errors as described by Equation (4-14). Substitutions are conveniently provided by Equations (4-15) and (4-16), respectively, for coherent and random amplitude error.

$$\varepsilon_{channel} = \overline{\varepsilon}_{sensor} + \overline{\varepsilon}_{filter} + \left[\varepsilon^2_{amplifier} + \varepsilon^2_{random} + \varepsilon^2_{coherent} \right]^{1/2} \qquad (4\text{-}14)$$

$$\varepsilon_{\text{coherent}} = \frac{V_{\text{cm}}}{V_{\text{diff}}} \cdot \left[\frac{R_{\text{diff}}}{R_{\text{cm}}}\right]^{1/2} \cdot \frac{Av_{\text{cm}}}{Av_{\text{diff}}} \cdot \left[1 + \left(\frac{f_{\text{coh}}}{f_c}\right)^{2n}\right]^{-1/2} \cdot 100\% \quad (4\text{-}15)$$

$$\varepsilon_{\text{random}} = \frac{V_{\text{cm}}}{V_{\text{diff}}} \cdot \left[\frac{R_{\text{diff}}}{R_{\text{cm}}}\right]^{1/2} \cdot \frac{Av_{\text{cm}}}{Av_{\text{diff}}} \cdot \left[\frac{2}{k}\left(\frac{f_c}{f_{hi}}\right)\right]^{1/2} \cdot 100\% \quad (4\text{-}16)$$

4-3 DC, SINUSOIDAL, AND HARMONIC SIGNAL CONDITIONING

Signal conditioning is concerned with upgrading the quality of a signal to the accuracy of interest coincident with signal acquisition, scaling, and band limiting. The unique requirements of each analog data acquisition channel plus the economic constraint of achieving only the performance necessary in specific applications are an impediment to standardized designs. The purpose of this chapter, therefore, is to develop a unified, quantitative design approach for signal acquisition and conditioning that offers new understanding and accountability measures. The following examples include both device and system errors in the evaluation of total signal conditioning channel error.

A dc and sinusoidal signal conditioning channel is considered that has widespread industrial application in process control and data logging systems. Temperature measurement employing a Type-C thermocouple is to be implemented over the range of 0 to 1800°C while attenuating ground-conductive and electromagnetically coupled interference. A 1-Hz signal BW is coordinated with filter cutoff to minimize the error provided by a single-pole filter, as described in Table 3-5 (see Chapter 3). Narrow-band signal conditioning is accordingly required for the differential-input 17.2 µV/°C thermocouple signal range of 0–31 mV dc, and for rejecting 1 V rms of 60 Hz common-mode interference, providing a residual coherent error of 0.009%FS. An OP-07A subtractor instrumentation-amplifier circuit combining a 22 Hz differential-lag RC low-pass filter is capable of meeting these requirements, including a full-scale output signal of 4.096 V dc with a differential gain $A_{v\text{diff}}$ of 132, without the cost of a separate active filter.

This austere dc and sinusoidal circuit is shown by Figure 4-5, with its parameters and defined-error performance tabulated in Tables 4-3 through 4-5. This $A_{v\text{diff}}$ further results in a −3dB frequency response of 4.5 kHz to provide a sensor-loop internal noise contribution of 4.4 μV_{pp} with 100 ohms source resistance. With 1% tolerance resistors, the subtractor amplifier presents a common-mode gain of 0.02 by the considerations of Table 2-2. The OP-07A error budget

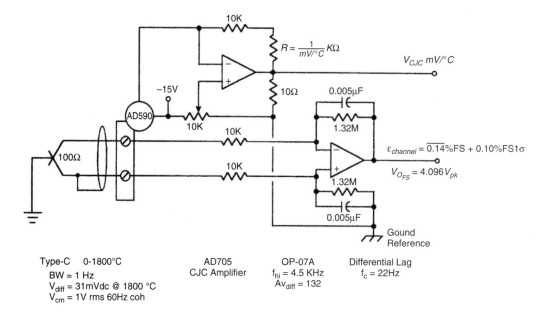

Figure 4-5. dc and ac sinusoidal signal conditioning.

of 0.103%FS is combined with other channel error contributions including a mean filter error of $\overline{0.1}$%FS and $\overline{0.011}$%FS linearized thermocouple. The total channel error of 0.246%FS at 1σ expressed in Table 4-5 is dominated by static mean error, which is an inflexible error to be minimized throughout all instrumentation systems. Postconditioning linearization software achieves a residual deviation from true temperature values of 0.2°C over 1800°C, and active cold-junction compensation of ambient temperature is provided by an AD590 sensor attached to the input terminal strip to within $\overline{0.5}$°C. Note that R_i is 10 K ohms.

Table 4-2. Signal bandwidth requirements

Signal	BW (Hz)
dc	$dV_s/\pi V_{FS}\, dt$
Sinusoidal	1/period T
Harmonic	10/period T
Single event	2/width τ

Table 4-3. Amplifier input parameters

Symbol	OP-07A	AD624C	AD215BY	Comment
V_{OS}	10 μV	25 μV	0.4 mV	Offset voltage
$\dfrac{dV_{OS}}{dT}$	0.2 μV/°C	0.25 μV/°C	2 μV/°C	Voltage drift
I_{OS}	0.3 nA	10nA	300nA	Offset current
$\dfrac{dI_{OS}}{dT}$	5 pA/°C	20 pA/°C	1 nA/°C	Current drift
$A_{v\text{diff}}$	132	50	1	Differential gain
$A_{v\text{cm}}$	0.02 (1%R)	0.0001	0.0001	Common-mode gain
CMRR	6600	5×10^5	10^4	$A_{v\text{diff}}/A_{v\text{cm}}$
CMV	10 V_{rms}	10 V_{rms}	1500 V_{rms}	Max applied volts
V_{Npp}	$6.6[(V_t^2 + V_n^2)f_{hi}]^{1/2}$	$6.6[(V_t^2 + V_c^2 + V_n^2)f_{hi}]^{1/2}$	$6.6[V_t^2 f_{hi}]^{1/2}$	Total input noise
V_t rms	1.3 nV/$\sqrt{\text{Hz}}$	4 nV/$\sqrt{\text{Hz}}$	0.9 nV/$\sqrt{\text{Hz}}$	Thermal noise
V_c rms	None	1.8 nV/$\sqrt{\text{Hz}}$	Negligible	Contact noise
V_n rms	10 nV/$\sqrt{\text{Hz}}$	4 nV/$\sqrt{\text{Hz}}$	Negligible	Amplifier noise
f_{hi}	4.5 KHz	150 KHz	120 KHz	–3db bandwidth
f_{contact}	None	100 Hz	100 Hz	Contact noise frequency
$\dfrac{dA_V}{dT}$	50 ppm/°C	5 ppm/°C	15 ppm/°C	Gain drift
$f(A_V)$	0.01%	0.001%	0.005%	Gain nonlinearity
R_{diff}	8×10^7 Ω	10^9 Ω	10^{12} Ω	Differential resistance
R_{cm}	2×10^{11} Ω	10^9 Ω	5×10^9 Ω	Common-mode resistance
R_s	100 Ω	1 K	50 Ω	Source resistance
V_{OFS}	4.096 V_{pk}	±5 V_{pp}	±5 V_{pp}	Full-scale output
dT	10°C	10°C	10°C	Temperature variation

The information content of instrumentation signals is described by their amplitude variation with time, or through Fourier transformation by signal bandwidth (BW) in Hz. Instrumentation signal types are accordingly classified in Table 4-2 with their minimum BW requirements specified in terms of signal waveform parameters. DC signal time rate of change is equated to the time derivative of a sinusoidal signal and evaluated at zero time to determine its BW requirement. In the case of harmonic signals, a first-order rolloff of –20dB/decade is assumed from a full-scale signal amplitude at the inverse-waveform period $1/T$, defining the fundamental frequency, declining to one-tenth of full scale at a BW value of ten times the fundamental frequency.

dc and sinusoidal channel	Harmonic channel
Sensor	
Type-C thermocouple 17.2 μV/°C post-conditioning linearization software $\dfrac{\overline{0.2°C}}{1800°C} \cdot 100\% = \overline{0.011}\%$ FS	1 KΩ piezoresistor bridge with $F = ma$ response 0.1%FS
Interface	
AD 590 temperature sensor cold-junction compensation $\dfrac{\overline{0.5°C}}{1800°C} \cdot 100\% = \overline{0.032}\%$FS	Regulators for sensor excitation ± 0.5 V dc ± 50 μV or 0.01%FS
Signal quality	

$$\varepsilon_{coh} = \frac{V_{cm}}{V_{diff}} \cdot \left[\frac{R_{diff}}{R_{cm}}\right]^{1/2} \cdot \frac{A_{V_{cm}}}{A_{V_{diff}}}$$

$$\cdot \left[1 + \left(\frac{f_{coh}}{f_c}\right)^{2n}\right]^{-1/2} \cdot 100\% \qquad (4\text{-}15)$$

$$= \frac{(1 \text{ V}_{rms}\, 2\sqrt{2})_{pp}}{31 \text{ mV}_{dc}} \cdot \left[\frac{80 \text{ M}\Omega}{200 \text{ G}\Omega}\right]^{1/2}$$

$$\cdot \frac{0.02}{132} \cdot \left[1 + \left(\frac{60 \text{ Hz}}{22 \text{ Hz}}\right)^2\right]^{-1/2} \cdot 100\%$$

$$= 0.009\%\text{FS}$$

$$\varepsilon_{rand} = \frac{V_{cm}}{V_{diff}} \cdot \left[\frac{R_{diff}}{R_{cm}}\right]^{1/2} \cdot \frac{A_{V_{cm}}}{A_{V_{diff}}}$$

$$\cdot \left[\frac{2}{K}\frac{f_c}{f_{hi}}\right]^{1/2} \cdot 100\% \qquad (4\text{-}16)$$

$$= \frac{1 \text{ V}}{7 \text{ mV}} \cdot \left[\frac{1 \text{ G}\Omega}{1 \text{ G}\Omega}\right]^{1/2} \cdot \frac{10^{-4}}{50}$$

$$\cdot \left[\frac{2}{0.9}\frac{3 \text{ kHz}}{150 \text{ kHz}}\right]^{1/2} \bullet 100\%$$

$$= 0.006\%\text{FS}$$

Considered now is the premium harmonic signal-conditioning channel of Figure 4-6, employing a 0.1%FS systematic error piezoresistive sensor that can transduce acceleration signals in response to applied mechanical force. A harmonic-signal spectral bandwidth is allowed for this example from dc to 1 KHz with the 1K-source-resistance bridge sensor generating a maximum input signal amplitude of 70 mV rms, up to 100 Hz fundamental frequencies, with rolloff at −20 db per decade of frequency to 7 mV rms at 1 KHz BW. The ±0.5 V dc bipolar sensor excitation is furnished by isolated three-terminal regulators to within ±50 μV dc variation, providing a negligible 0.01%FS differential-mode error. The

Table 4-4. Amplifier error budgets

$\varepsilon_{amplRTI}$	OP-07A	AD624C	AD215BY
V_{os}	$\overline{10}\ \mu V$	Trimmed	Timmed
$\dfrac{dV_{os}}{dT} \cdot dT$	$2\ \mu V$	$2.5\ \mu V$	$20\ \mu V$
$I_{os} \cdot R_i$	$\overline{3}\ \mu V$	$\overline{10}\ \mu V$	$\overline{15}\ \mu V$
V_{Npp}	$4.4\ \mu V$	$15\ \mu V$	$2\ \mu V$
$f(A_V) \cdot \dfrac{V_{O_{FS}}}{A_{V_{diff}}}$	$\overline{3}\ \mu V$	$\overline{1}\ \mu V$	$\overline{250}\ \mu V$
$\dfrac{dA_V}{dT} \cdot dT \cdot \dfrac{V_{O_{FS}}}{A_{V_{diff}}}$	$15.5\ \mu V$	$5\ \mu V$	$750\ \mu V$
$\Sigma \overline{\text{mean}} + 1\sigma\,\text{RSS}$	$(\overline{16} + 16)\ \mu V$	$(\overline{11} + 16)\ \mu V$	$(\overline{265} + 750)\ \mu V$
$X \dfrac{A_{V_{diff}}}{V_{O_{FS}}} \cdot 100\%$	$0.103\%\text{FS}$	$0.027\%\text{FS}$	$0.020\%\text{FS}$

sensor-shield-buffered common-mode-voltage active drive also preserves signal conditioning CMRR over extended cable lengths.

An AD624C preamplifier raises the differential sensor signal to a $\pm 5\ V_{pp}$ full-scale value while attenuating 1 V rms of common-mode random interference, in concert with the low-pass filter, to a residual error of 0.006%FS, as defined by Equation (4-16). The error budgets of the preamplifier and isolation

Table 4-5. Signal conditioning channel error summary

Element	DC Sinusoidal $\varepsilon_{\%FS}$	Comment	Harmonic $\varepsilon_{\%FS}$	Comment
Sensor	$\overline{0.011}$	Type-C linearized	0.100	Piezoresistor
Interface	$\overline{0.032}$	CJC sensor	0.010	Sensor excitation
Amplifier	0.103	OP-07A	0.027	AD624C
Isolator	None		0.020	AD215AY
Filter	$\overline{0.100}$	Table 3-5	$\overline{0.115}$	Table 3-6
Signal quality	0.009	60 Hz ε_{coh}	0.006	Noise ε_{rand}
$\varepsilon_{channel}$	$\overline{0.143}\%\text{FS}$	$\Sigma\overline{\text{mean}}$	$\overline{0.115}\%\text{FS}$	$\Sigma\overline{\text{mean}}$
	$0.103\%\text{FS}$	$1\sigma\,\text{RSS}$	$0.106\%\text{FS}$	$1\sigma\,\text{RSS}$
	$0.246\%\text{FS}$	$\Sigma\overline{\text{mean}} + 1\sigma\,\text{RSS}$	$0.221\%\text{FS}$	$\Sigma\overline{\text{mean}} + 1\sigma\,\text{RSS}$
	$0.761\%\text{FS}$	$\Sigma\overline{\text{mean}} + 6\sigma\,\text{RSS}$	$0.751\%\text{FS}$	$\Sigma\overline{\text{mean}} + 6\sigma\,\text{RSS}$

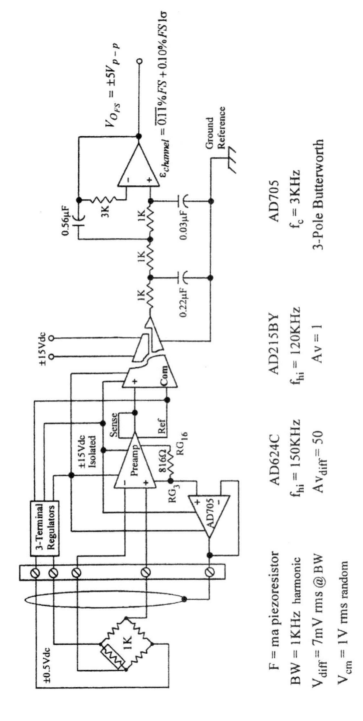

Figure 4-6. Premium harmonic signal conditioning.

100

$V_{O_{FS}} = \pm 5V_{p-p}$

$\varepsilon_{channel} = \overline{0.11\% FS} + 0.10\% FS 1\sigma$

Ground
Reference

0.56μF

3K

1K

1K

1K

0.03μF

0.22μF

AD705
$f_c = 3KHz$
3-Pole Butterworth

±15Vdc

AD215BY
$f_{hi} = 120KHz$
$Av = 1$

Sense

Ref

Com

±15Vdc
Isolated

Preamp

816Ω

RG_{16}

RG_3

AD705

AD624C
$f_{hi} = 150KHz$
$Av_{diff} = 50$

3-Terminal
Regulators

1K

±0.5Vdc

F = ma piezoresistor
BW = 1KHz harmonic
$V_{diff} = 7mV$ rms @BW
$V_{cm} = 1V$ rms random

amplifier, tabulated in Tables 4-3 and 4-4, also include a sensor-loop internal noise contribution of 15 μV_{pp} based on the provisions of Figure 4-2, where the $1/f$ contact noise frequency is taken as 10% of signal BW. Three contributions comprising this internal noise are evaluated as source resistance thermal noise V_t, contact noise V_c arising from 1 mA of dc current flow, and amplifier internal noise V_n. The three-pole Butterworth low-pass filter cutoff frequency is derated to a value of 3 BW to minimize its device error. Note that the AD705 filter amplifier is included in the mean filter device error of $\overline{0.115}$%FS. The total channel 1σ instrumentation error of 0.221%FS consists of an approximate equal sum of static mean and variable systematic error values at 1σ confidence in Table 4-5.

4-4   ANALOG SIGNAL PROCESSING

When achievable analog signal conditioning error does not meet minimum measurement requirements, identical channels may be averaged to reduce the total error. Random and systematic errors added to the value of a measurement can be reduced by taking the arithmetic mean of a sum of n independent measurement values. This assumes that combined systematic error contributions are sufficient in number to approximate a zero mean value, and likewise for random errors. Sensor device error is frequently simplified in its specification as the nonlinearity of its transfer function and conservatively represented by a mean error. However, many effects actually contribute to sensor error, such as material–energy interactions, which are unknown other than their dependence on random variables that generally are compliant to reduction by arithmetic-mean averaging.

The foregoing conditions are sufficiently met by typical signal conditioning channels to enable averaged outputs consisting of arithmetic signal additions and RSS error additions. This provides signal quality improvement by $n/\sqrt{n}$ and channel error reduction by its inverse. Averaged measurement error accordingly corresponds to the error of any one identical channel divided by $\sqrt{n}$. However, diminishing returns may result in an economic penalty to achieve error reduction beyond a few channels combined. Further, signal conditioning mean filter device error also remains additive, which is a limitation remedied by relocating the channel filter postaveraging.

Figure 4-7 describes signal conditioning channel averaging in which amplifier stacking between respective device outputs and ground provide arithmetic sig-

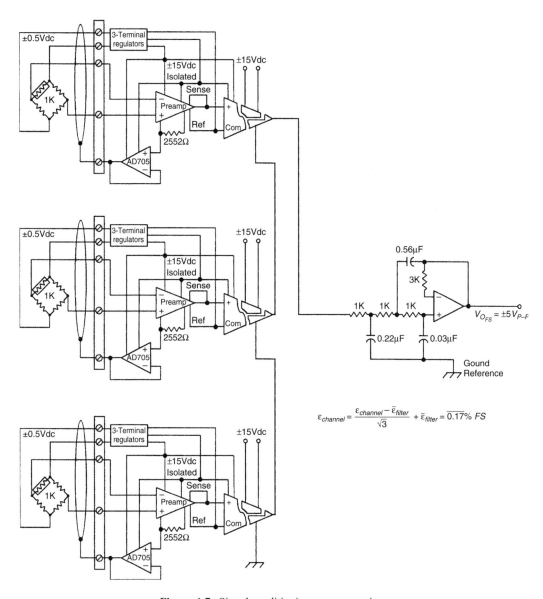

$$\varepsilon_{channel} = \frac{\varepsilon_{channel} - \overline{\varepsilon}_{filter}}{\sqrt{3}} + \overline{\varepsilon}_{filter} = \overline{0.17}\% \ FS$$

Figure 4-7. Signal conditioning error averaging.

nal additions, and their parallel inputs provide RSS error additions. The $A_{v\text{diff}}$ values of each stage are equally scaled so that the sum of n outputs achieves the full-scale value for a single channel. The three averaged harmonic signal conditioning channels, therefore, each require an $A_{v\text{diff}}$ of 16.67 for a per-channel output of 1.667 V by employing gain resistors of 2552 Ω. With reference to Table 4-5, moving the filter post-averaging provides an improved overall error of $(0.221\% - \overline{0.115}\%)/\sqrt{3} + \overline{0.115}\%$, approximately totaling $\overline{0.176}\%$FS. Note that this connection obviates the requirement for an output summing amplifier and its additional device error contribution.

The utility of logarithmic functions in signal processing primarily lies in their ability to accommodate wide-dynamic-range signals. Logarithmic arguments are always dimensionless. Consequently, the log of the ratio of two voltages or currents are required in logarithmic realizations, with the denominator term normally a fixed reference value, as described in Table 4-6. Commercially available log devices offer performance extending over four voltage decades, typically from 1 mV to 10 V, and six current decades from 1 nA to 1 mA. The bipolar log function is especially useful and is symmetrical about its linear segment centered on zero input signals, as shown in Figure 4-8.

The application of a logarithmic function at the input of a data acquisition system has merit for processing high-resolution sensor data by compressing an

Table 4-6. Logarithmic functions

Function	Description
$V_0 = C \cdot \log_{10}\left(\dfrac{V_i}{V_{\text{ref}}}\right)$	Voltage log
$V_0 = C \cdot \log_{10}\left(\dfrac{I_i}{I_{\text{ref}}}\right)$	Current log
$V_0 = V_{\text{ref}} \cdot 10^{-V_i/C}$	Antilog
$V_0 = C \cdot \log 10\left(\dfrac{V_1}{V_2}\right)$	Log ratio
$V_0 = C \cdot \sinh^{-1}\left(\dfrac{V_i}{2V_{\text{ref}}}\right)$	Bipolar log
$V_0 = C \cdot \sinh\left(\dfrac{V_i}{V_{\text{ref}}}\right)$	Bipolar antilog

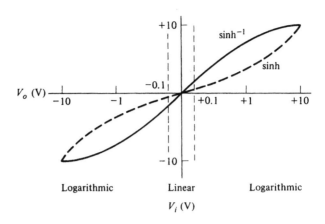

Figure 4-8. Bipolar log and antilog functions.

input signal to conform to a limited system dynamic range. This can be appreciated from observation of Figure 4-8, where it is apparent that the logarithmic gain is unity for a 10-V full-scale input signal, and logarithmically increases to a maximum gain at minimum input levels. However, a mirror-symmetry antilog operation is required following output signal recovery for linear signal representation. This input compression and output expansion, or companding operation, is bene-

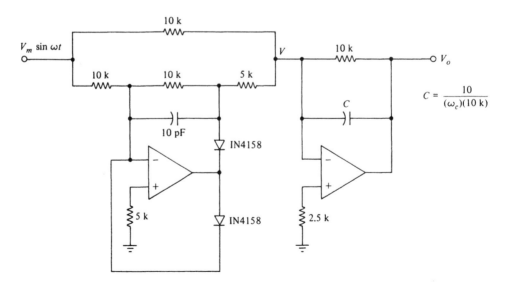

Figure 4-9. Precision ac-to-dc converter.

ficial in extending the dynamic range of truncated-word-length digital processors by effectively increasing the number of A/D converter quantizing levels at lower signal amplitudes. The typical log conformity error is 1% of full scale, and this constant fractional error is maintained throughout the signal dynamic range in logarithmic form at the expense of high signal resolution at any point within the range.

An occasionally required signal processing operation is the precision rectification of very low-level signals. Passive full-wave rectification is inadequate for this task because silicon diodes will not conduct until an applied voltage exceeds approximately 600 mV. However, the active full-wave rectification and smoothing circuit of Figure 4-9 will provide accurate ac-to-dc conversion down to submillivolt signal levels. The circuit shown generates a conversion error of 0.6%FS, determined primarily by the residual harmonic distortion passed by the RC smoothing filter with a cutoff frequency of one-tenth the input signal frequency.

For critical measurements, the dual redundant signal conditioning channels of Figure 4-10 provide continuous diagnostics for failures in either channel [8]. Channel states are illustrated by the decision tree representing channel a-priori

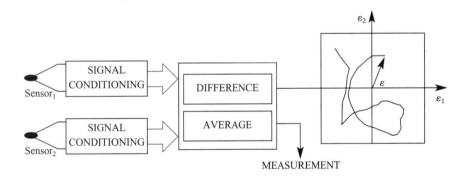

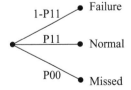

Figure 4-10. Redundant sensor diagnostics.

probabilities. Further, averaging two channels provides a beneficial measurement error reduction to 0.707 of the error for a single channel. Identical sensor channels achieve effective failure detection when the difference vector of their signals exceeds the channel signal conditioning error window limits shown. This method employs modeled per-channel error developed in this chapter, and utilizes continuous sensor measurement values rather than comparisons of measurement values with independent analytical predictions. A failure of either channel, denoted by (1-P11), immediately exceeds error window limits. Alternatively, unlikely missed detection would arise from simultaneous sensor channel failures, denoted by P00, with the inability for their signal difference to exceed error window limits.

PROBLEMS

4-1. The universal signal conditioning channel illustrated in Figure 4-11 provides transient-free shunt resistor gain switching of $\times 1$, $\times 10$, $\times 100$, or $\times 500$ based upon AD624 amplifier external R_g resistance yielding an $Av = (40,000/R_g) + 1$. This is followed by a three-pole selectable Bessel or Butterworth low-pass signal filter, also with a shunt resistor switching for f_c corner frequencies of 10 Hz, 100 Hz, or 1 KHz. Determine the correspond-

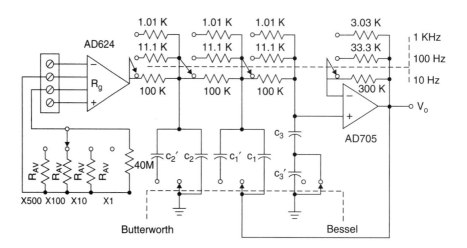

Figure 4-11.

ing gain selection R_{Av} resistor values, and both Bessel and Butterworth C_1, C_2, and C_3 capacitor values that satisfy the circuit shown. Show calculations.

4-2. A subtractor circuit with one-pole bandlimiting filter (Figure 4-12) provides an austere signal-conditioning channel that is to be evaluated with an OP-07A amplifier for an input dc V_{diff} of 10 mV raised to a full-scale V_0 of 1 V. Consider a signal spectral occupancy requirement of one-half Hz, and 1 V_{rms} of both random and 60 Hz coherent common-mode interference to be upgraded in quality at the amplifier output. Amplifier input noise includes device V_n and thermal noise V_t contributions considering an R_s of 1 K. Evaluate the channel combined device and signal 1σ errors for a temperature variation of 10°C and the use of 1% tolerance resistors with a 50 ppm per °C temperature coefficient. Assume amplifier V_{OS} is trimmed to zero. Show all calculations and tabulate results.

4-3. Verify the input noise values for the AD624C amplifier tabulated in Table 4-3, including calculation of the total peak-to-peak value. The constant 6.6 in Equation (4-3) provides the crest factor of rms Gaussian noise in its conversion to peak to peak, within 0.1% error, compatible for combining with other amplifier error parameter dimensional values.

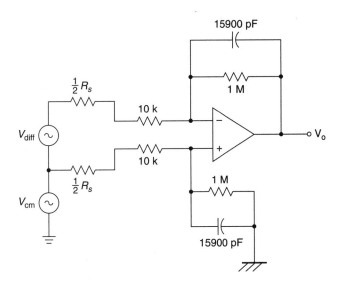

Figure 4-12.

BIBLIOGRAPHY

1. Akao, Y., "Quality Function Deployment and CWQC in Japan," *Quality Progress,* October 1983, pp. 25–29.
2. Budai, M., "Optimization of the Signal Conditional Channel," Senior Design Project, *Electrical Engineering Technology,* University of Cincinnati, 1978.
3. *Designers Reference Manual,* Analog Devices, Norwood, MA, 1996.
4. Fano, R.M., "Signal to Noise Ratio in Correlation Detectors," *MIT Technical Report 186,* 1951.
5. Garrett, P.H., *Analog Systems for Microprocessors and Minicomputers,* Reston, 1978.
6. Garrett, P.H., *High Performance Instrumentation and Automation,* CRC Press, 2005.
7. Gordon, B.M., *The Analogic Data-Conversion Systems Digest,* Analogic, Wakefield, MA, 1977.
8. Hunt, V.J., *Redundant Structures for Fault-Tolerant Control,* M.S. Thesis, Elect. and Comp. Engr., Univ. of Cincinnati, 1991.
9. Ott, H.W., *Noise Reduction Techniques in Electronic Systems,* Wiley-Interscience, 1976.
10. Petriu, E.M., Ed., *Instrumentation and Measurement Technology and Applications,* IEEE Selected Conference Papers, ISBN 0-7803-4268-2, 1998.
11. Raemer, H.R., *Statistical Communications Theory and Appications,* Prentice-Hall, 1969.
12. Schwartz, M., Bennett, W., and Stein, S., *Communications Systems and Techniques,* McGraw-Hill, 1966.
13. Sheingold, D.H., Ed., Transducer Interfacing Handbook, Analog Devices, Norwood, MA, 1980.
14. Zuch, E., *Data Acquisition and Conversion Handbook,* Datel-Intersil, Mansfield, MA, 1982.

CHAPTER 5

Data Converstion Devices and Parameters

5-0   INTRODUCTION

Data-conversion devices provide the interfacing components between continuous-time signals representing the parameters of physical processes and their discrete-time digital equivalent. Recent emphasis on computer systems for automated manufacturing and the growing interest in using personal computers for data acquisition and control have increased the need for improved understanding of the design requirements of real-time computer I/O systems. However, before describing the theory and practice involved in these systems it is advantageous to understand the characterization and operation of the various devices from which these systems are fabricated. This chapter accordingly provides detailed information concerning A/D and D/A data conversion devices, including seven application-specific A/D converters, with supporting components including analog multiplexers and sample-hold devices. The development of the individual error budgets representing these devices is also provided to continue the quantitative methodology of this text.

5-1   ANALOG MULTIPLEXERS

Field-effect transistors (FETs), both CMOS and JFET, are universally used as electronic multiplexer switches today, displacing earlier bipolar devices that had volt-

Advanced Instrumentation and Computer I/O Design, Second Edition. By Patrick H. Garrett
Copyright © 2013 the Institute of Electrical and Electronics Engineers, Inc.

Table 5-1. Multiplexer switch characteristics

Type	ON resistance	OFF isolation	Sample rate
CMOS	100 Ω	70 dB	10 MHz
JFET	50 Ω	70 dB	1 MHz
Reed	0.1 Ω	90 dB	1 KHz

age offset problems. Junction FET switches have greater device electrical ruggedness and approximately the same switching speeds as CMOS devices. However, CMOS switches are dominant in multiplexer applications because of their unfailing turnoff, especially when the power is removed, unlike JFET devices, and their ability to multiplex signal levels up to the power supply voltages. Figure 5-1 describes a CMOS analog switch circuit in which a stable ON resistance is achieved of about 100 Ω series resistance by the parallel p- and n-channel devices. Terminating a CMOS multiplexer with a high-input-impedance voltage follower eliminates any voltage-divider errors possible as a consequence of the ON resistance. Figure 5-2 presents interconnection configurations for a multiplexer.

Errors associated with analog multiplexers are tabulated in Table 5-2, and are dominated by the average transfer error defined by Equation (5-1). This error is essentially determined by the input-voltage-divider effect, and is minimized to a typical value of $\overline{0.01}$%FS when the AMUX is followed by an output buffer amplifier. The input amplifier associated with a sample-hold device often provides this high-impedance termination. Another error that can be significant is OFF-channel leakage current that creates an offset voltage across the input source resistance.

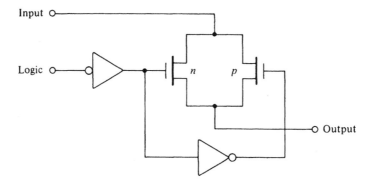

Figure 5-1. CMOS analog switch.

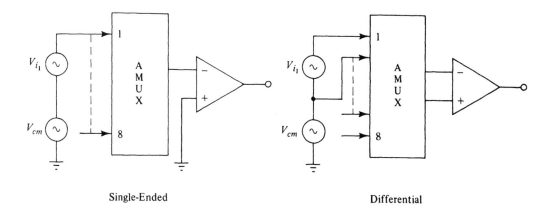

Single-Ended   Differential

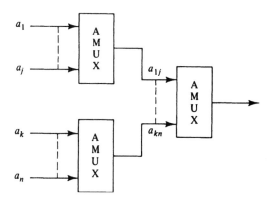

Figure 5-2. Multiplexer interconnections and array.

Table 5-2. Representative multiplexer errors

Parameter		Reed	CMOS
Transfer error		$\overline{0.01}$%	$\overline{0.01}$%
Crosstalk error		0.001%	0.001%
Leakage error			0.001%
Thermal offset		0.001%	
ε_{AMUX}	$\Sigma\overline{mean} + 1\ \sigma$ RSS	$\overline{0.01}$%FS	$\overline{0.01}$%FS

$$\text{Transfer error} = \frac{V_i - V_o}{V_i} \times 100\% \qquad (5\text{-}1)$$

5-2   SAMPLE-HOLD DEVICES

Sample-hold (S/H) devices provide an analog signal memory function for use in sampled-data systems for temporary storage of changing signals for data conversion purposes. Sample holds are available in several circuit variations, each suited to specific speed and accuracy requirements. Figure 5-3 describes a contemporary circuit that may be optimized either for speed or accuracy. The noninverting input amplifier provides a high-impedance buffer stage, and the overall unity feedback minimizes signal transfer error when the device is in the tracking mode. The clamping diodes insure that the circuit remains stable during the hold mode when the switch is open. The inclusion of S/H devices in sampled-data systems must be carefully considered. The following examples represent the three essential applications for sample holds.

Figure 5-4 diagrams a conventional multiplexed data conversion system cycle. The multiplexer and external circuit of Channel 1 are sampled by the S/H for a time sufficient for signal settling to within the amplitude error of interest. For sensor channels having RC time constants on the order of the S/H internal acquisition time, defined by Equation (5-2), overlapping multiplexer channel selection and A/D conversion can speed system throughput significantly by means of an interposed sample hold. A second application is described by Figure 5-5. Simultaneous data acquisition is required for many laboratory measurements in which

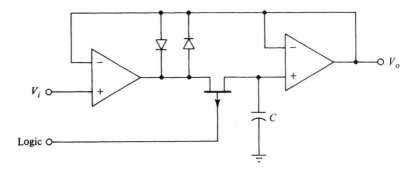

Figure 5-3. Closed-loop sample hold.

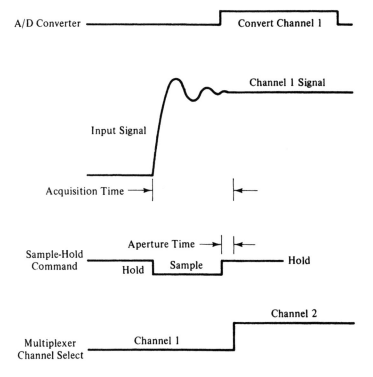

Figure 5-4. Multiplexed conversion system timing diagram.

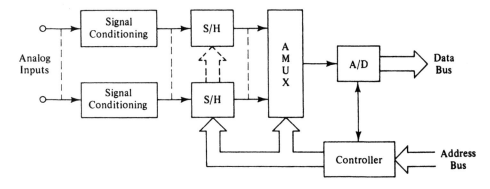

Figure 5-5. Simultaneous data acquisition.

multiple sensor channels must be acquired at precisely the same time. By matching S/H devices in bandwidth and aperture time, interchannel signal time skew can be minimized. The timing relationships are consequently preserved between signals even though data conversion is performed sequentially.

$$\text{Acquisition time} = \frac{|V_o - V_i|}{I_o} + 9(R_o + R_{on})C \text{ seconds} \qquad (5\text{-}2)$$

Voltage-comparison A/D converters such as successive approximation devices require a constant signal value for accurate conversion. This function is normally provided by the application of a sample hold preceding the A/D converter, which constitutes the third application. An important issue is matching of S/H and A/D specifications to achieve the performance of interest. Sample-hold performance is principally determined by the input amplifier bandwidth and current output capability, which determines its ability to drive the hold capacitor C. A limiting parameter is the acquisition time of Equation (5-2) and Figure 5-6 which when added to the conversion period T of an A/D converter determines the maximum throughput performance possible for a S/H and connected A/D. As a specific example, an Analog Devices 9100 device has an acquisition time of 16 ns for 0.01%FS (13-bit) settling, enabling data conversion rates $1/(16\text{ ns} + T)$ Hz. In the

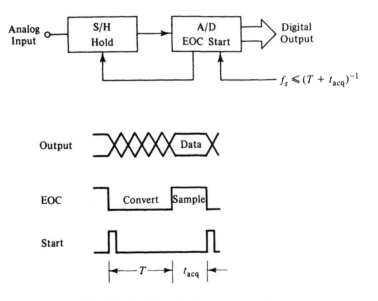

Figure 5-6. S/H–A/D timing relationships.

Table 5-3. Representative sample-hold errors

Acquisition error		0.01%
Nonlinearity		0.004%
Gain		0.01%
Temperature coefficient		0.001%
$\varepsilon_{S/H}$	$\Sigma\overline{\text{mean}} + 1\ \sigma$ RSS	0.02%FS

sample mode, the charge on the hold capacitor is initially changed at the slew-limited output current capability I_o of the input amplifier. As the capacitor voltage enters the settling band coincident with the linear region of amplifier operation, final charging is exponential and corresponds to the summed time constants in Equation (5-2), where R_o corresponds to amplifier output resistance and R_{ON} the switch resistance. The consequence of aperture time is to provide an average aperture error associated with the finite bound within which the amplitude of a sampled signal is acquired. Since this is a system error, instead of a component error, its evaluation is deferred until Section 6-3.

5-3   DIGITAL-TO-ANALOG CONVERTERS

Digital-to-analog (D/A) converters, or DACs, provide reconstruction of discrete-time digital signals into continuous-time analog signals for computer-interfacing output-data recovery purposes such as actuators, displays, and signal synthesizers. D/A converters are considered prior to analog-to-digital (A/D) converters because some A/D circuits require DACs in their implementation. A D/A converter may be considered a digitally controlled potentiometer that provides an output voltage or current normalized to a full-scale reference value. A descriptive way of indicating the relationship between analog and digital conversion quantities is a graphical representation. Figure 5-7 describes a three-bit D/A converter transfer relationship having eight analog output levels ranging between zero and seven-eighths of full scale. Notice that a DAC full-scale digital input code produces an analog output equivalent to FS – 1 LSB. The basic structure of a conventional D/A converter includes a network of switched current sources having MSB to LSB values according to the resolution to be presented. Each switch closure adds a binary-weighted current increment to the output bus. These current contributions are then summed by a current-to-voltage converter amplifier in a manner appropriate to scale the output signal. Figure 5-8 illustrates such a structure for a three-bit DAC with unipolar straight binary coding corresponding to the representation of Figure 5-7.

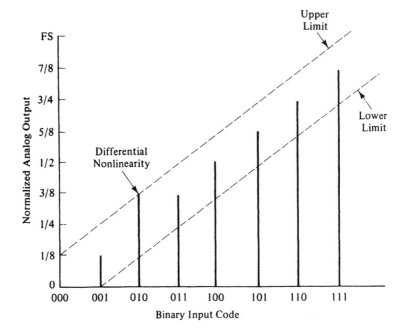

Figure 5-7. Three-bit D/A converter relationships.

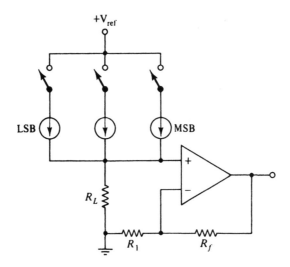

Figure 5-8. Straight binary three-bit DAC.

In practice, the realization of the transfer characteristic of a D/A converter is nonideal. With reference to Figure 5-7, the zero output may be nonzero because of amplifier offset errors, the total output range from zero to FS – 1 LSB may have an overall increasing or decreasing departure from the true encoded values resulting from gain error, and differences in the height of the output bars may exhibit a curvature owing to converter nonlinearity. Gain and offset errors may be compensated for, leaving the residual temperature-drift variations shown in Table 5-4 as the temperature coefficient (tempco) of a representative 12-bit D/A converter. A voltage reference is necessary to establish a basis for the DAC absolute output voltage. The majority of voltage references utilize the bandgap principle, whereby the V_{be} of a silicon transistor has a negative tempco of –2 mV/°C that can be extrapolated to approximately 1.2 V at absolute zero (the bandgap voltage of silicon).

Converter nonlinearity is minimized through precision components because it is essentially distributed throughout the converter network and cannot be eliminated by adjustment as with gain and offset errors. Differential nonlinearity and its variation with temperature are prominent in data converters in that they describe the difference between the true and actual outputs for each of the 1 LSB code changes. A DAC with a 2 LSB output change for a 1 LSB input code change exhibits 1 LSB of differential nonlinearity as shown. Nonlinearities greater than 1 LSB make the converter output no longer single valued, in which case it is said to be nonmonotonic and to have missing codes. Integral nonlinearity is an average error that generally does not exceed 1 LSB of the converter resolution as the sum of differential nonlinearities.

Table 5-5 presents frequently applied unipolar and bipolar codes expressed in terms of a 12-bit binary word length. These codes are applicable to both D/A and A/D converters. The choice of a code should be appropriate to the application and its sense understood (positive-true, negative-true). Positive-true coding defines a logic 1 as the positive logic level, and in negative-true coding the negative logic level is 1 with the other level 0. All codes utilized with data converters are based on the binary number system. Any base 10 number may be represented by Equation (5-3), where the coefficient a_i assumes a value of 1 or 0 between the MSB

Table 5-4. Representative 12-bit DAC errors

Mean integral nonlinearity (1 LSB)	0.024%
Temperature coefficient (1 LSB)	0.024%
Noise + distortion	0.001%
$\varepsilon_{D/A}$ $\Sigma\overline{mean} + 1\ \sigma\,RSS$	0.048%FS

Table 5-5. Data converter binary codes

Unipolar codes—12-bit converters
Straight binary and complementary binary

Scale	+ 10 VFS	+ 5 VFS	Straight binary	Complementary binary
+ FS – 1 LSB	+ 9.9976	+ 4.9988	1111 1111 1111	0000 0000 0000
+ 7/8 FS	+ 8.7500	+ 4.3750	1110 0000 0000	0001 1111 1111
+ 3/4 FS	+ 7.5000	+ 3.7500	1100 0000 0000	0011 1111 1111
+ 5/8 FS	+ 6.2500	+ 3.1250	1010 0000 0000	0101 1111 1111
+ 1/2 FS	+ 5.0000	+ 2.5000	1000 0000 0000	0111 1111 1111
+ 3/8 FS	+ 3.7500	+ 1.8750	0110 0000 0000	1001 1111 1111
+ 1/4 FS	+ 2.5000	+ 1.2500	0100 0000 0000	1011 1111 1111
+ 1.8 FS	+ 1.2500	+ 0.6250	0010 0000 0000	1101 1111 1111
0 + 1 FS	+ 0.0024	+ 0.0012	0000 0000 0000	1111 1111 1110
0	0.0000	0.0000	0000 0000 0000	1111 1111 1111

BCD and complementary BCD

Scale	+ 10 VFS	+ 5 VFS	Binary coded decimal	Complementary binary coded decimal
+ FS – 1 LSB	+ 9.99	+ 4.95	1001 1001 1001	0110 0110 0110
+ 7/8 FS	+ 8.75	+ 4.37	1000 0111 0101	0111 1000 1010
+ 3/4 FS	+ 7.50	+ 3.75	0111 0101 0000	1000 1010 1111
+ 5/8 FS	+ 6.25	+ 3.12	0110 0010 0101	1001 1101 1010
+ 1/2 FS	+ 5.00	+ 2.50	0101 0000 0000	1010 1111 1111
+ 3/8 FS	+ 3.75	+ 1.87	0011 0111 0101	1100 1000 1010
+ 1/4 FS	+ 2.50	+ 1.25	0010 0101 0000	1101 1010 1111
+ 1.8 FS	+ 1.25	+ 0.62	0001 0010 0101	1110 1101 1010
0 + 1 FS	+ 0.01	+ 0.00	0000 0000 0001	1111 1111 1110
0	0.00	0.00	0000 0000 0000	1111 1111 1111

Bipolar codes—12-bit converters

Scale	± 5 VFS	Offset binary	Twos complement	Ones complement	Sign–Mag Binary
+ FS – 1 LSB	4.9976	1111 1111 1111	0111 1111 1111	0111 1111 1111	1111 1111 1111
+ 3/4 FS	+ 3.7500	1110 0000 0000	0110 0000 0000	0110 0000 0000	1110 0000 0000
+ 1/2 FS	+ 2.5000	1100 0000 0000	0100 0000 0000	0100 0000 0000	1100 0000 0000
+ 1/4 FS	+ 1.2500	1010 0000 0000	0010 0000 0000	0010 0000 0000	1010 0000 0000
0	0.0000	1000 0000 0000	0000 0000 0000	0000 0000 0000	1000 0000 0000
– 1/4 FS	– 1.2500	0110 0000 0000	1110 0000 0000	1101 1111 1111	0010 0000 0000
– 1/2 FS	– 2.5000	0100 0000 0000	1100 0000 0000	1011 1111 1111	0100 0000 0000
– 3/4 FS	– 3.7500	0010 0000 0000	1010 0000 0000	1001 1111 1111	0110 0000 0000
– FS + 1 LSB	– 4.9976	0000 0000 0001	1000 0000 0001	1000 0000 0000	0111 1111 1111
– FS	– 5.0000	0000 0000 0000	1000 0000 0000		

(0.5) and LSB (2^{-n}). This coding scheme is convenient for data converters in which the encoded value is interpreted in terms of a fraction of full scale for n-bit lengths. Straight-binary, positive-true unipolar coding is most commonly encountered. Complementary binary positive-true coding is identical to straight binary negative-true coding. Sign-magnitude bipolar coding is often used for outputs that are frequently in the vicinity of zero. Offset binary is readily converted to the more computer-compatible two-complement code by complementing the MSB.

$$N = \sum_{i=0}^{n} a_i \, 2^{-i} \tag{5-3}$$

As the input code to a DAC is increased or decreased it passes through major and minor transitions. A major transition is at half-scale when the MSB is switched and all other switches change state. If some switched current sources lag others, then significant transient spikes are generated, known as glitches. Glitch energy is of concern in fast-switching DACs driven by high-speed logic with time skew between transitions. However, high-speed DACs also frequently employ an output S/H circuit to deglitch major transitions by remaining in the hold-mode during these intervals. Internally generated noise is usually not significant in D/A converters except at extreme resolutions, such as in the 20-bit Analog Devices DAC 1862, whose LSB is equal to 10 μV with 10 V_{FS} scaling.

The advent of monolithic D/A converters has resulted in almost universal acceptance of the R-2R network DAC because of the relative ease of achieving precise resistance ratios with monolithic technology. This is in contrast to the low yields experienced with achieving precise absolute resistance values required by weighted-resistor networks. Equations (5-4) and (5-5) define the quantities of each converter as illustrated in Figure 5-9. For the R-2R network, an effective resistance of 3 R is seen by V_{ref} for each branch connection with equal left–right current division.

$$V_o = \frac{R_f}{R} \cdot V_{\text{ref}} \cdot \sum_{i=0}^{n} 2^{-i} \qquad \text{weighted resistor} \tag{5-4}$$

$$V_o = \frac{R_f}{2R} \cdot \frac{V_{\text{ref}}}{3} \cdot \sum_{i=0}^{n} 2^{-i} \qquad \text{R-2R resistor} \tag{5-5}$$

A D/A converter that accepts a variable reference can be configured as a multiplying DAC that is useful for many applications requiring a digitally controlled scale factor. Both linear and logarithmic scale factors are available for applica-

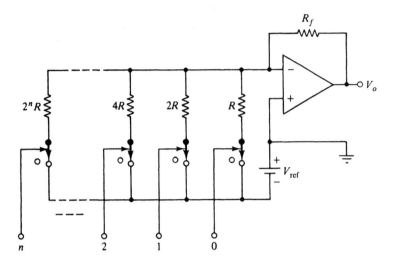

Figure 5-9a. Weighted resistor D/A converter.

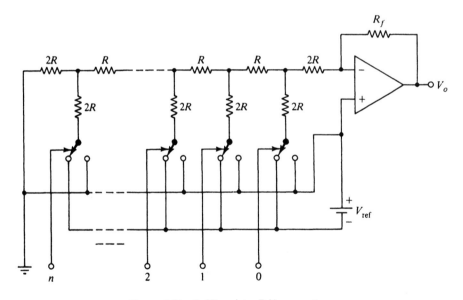

Figure 5-9b. R-2R resistor D/A converter.

tions such as, respectively, digital excitation in test systems and a dB step attenuator in communications systems. The simplest devices operate in one quadrant with a unipolar reference signal and digital code. Two-quadrant multiplying DACs utilize either bipolar reference signals or bipolar digital codes. Four-quadrant multiplication involves both a bipolar reference signal and bipolar digital code. Table 5-6 describes a two-quadrant, 12-bit linear multiplying D/A converter. The variable transconductance property made possible by multiplication is useful for many signal conditioning applications including programmable gain.

As system peripheral complexity has expanded to require more of a host computer's resources, peripheral interface devices have been provided with transparent processing capabilities to more efficiently distribute these tasks. In fact, some peripherals are more complicated than the host computer they support such as 4G communications processors. Universal peripheral bus master devices have evolved that offer a flexible combination of memory-mapped, interrupt-driven, and DSP capabilities with FIFO buffer memory for accommodation of multiple buses at differing speeds. The elementary D/A peripheral interface of Figure 5-10, in contrast, employs a program-initiated output whose status is polled by the host for a Ready enable. Data is then transferred to the D port with $\overline{\text{IOW}}$ low and $\overline{\text{CE}}$ high.

5-4   ANALOG-TO-DIGITAL CONVERTERS

The conversion of continuous-time analog signals to discrete-time digital signals is fundamental to obtaining a representative set of numbers that can be used by a

Table 5-6. Two-quadrant multiplying 12-bit DAC

Straight binary input	Analog output
1111 1111 1111	$\pm V_i \left(\dfrac{4095}{4096} \right)$
1000 0000 0001	$\pm V_i \left(\dfrac{2048}{4096} \right)$
0000 0000 0001	$\pm V_i \left(\dfrac{1}{4096} \right)$
0000 0000 0000	0 V

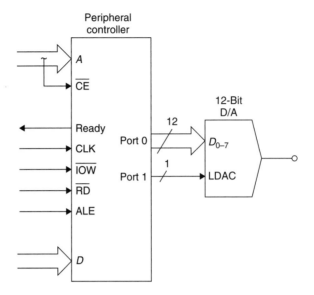

Figure 5-10. D/A peripheral interface.

digital computer. The three functions of sampling, quantizing, and encoding are involved in this process and implemented by all A/D converters, as illustrated by Figure 5-11. The detailed analytical considerations associated with these functions and their relationship to computer interface design are developed in Chapter 6. We are concerned here with A/D converter devices and their functional operations as we were with the previously described D/A conversion devices. In practice, one conversion is performed each period T, the inverse of sample rate f_s, whereby a numerical value derived from the converter quantizing levels is trans-

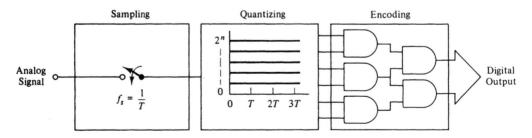

Figure 5-11. A/D converter functions.

lated to an appropriate output code. The graph of Figure 5-12 describes A/D converter input–output relationships and quantization error for prevailing uniform quantization, where each of the levels q is of spacing 2^{-n} (1 LSB) for a converter having an n-bit binary output word length. Note that the maximum output code does not correspond to a full-scale input value, but instead to $(1 - 2^{-n}) \cdot$ FS because there exist only $(2^n - 1)$ coding points, as shown in Figure 5-12.

Quantization of a sampled analog waveform involves the assignment of a finite number of amplitude levels corresponding to discrete values of input signal V_S between 0 and V_{FS}. The uniformly spaced quantization intervals 2^{-n} represent the resolution limit for an n-bit converter, which may also be expressed as the

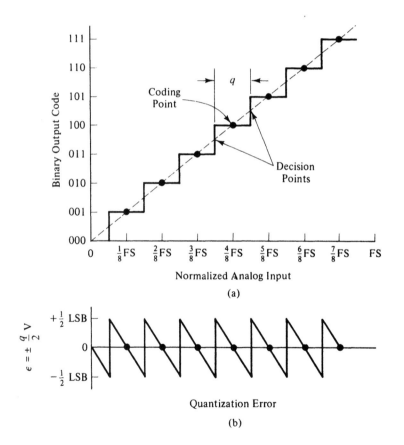

Figure 5-12. Three-bit A/D converter relationships: (a) quantization intervals, (b) quantization error.

quantizing interval q equal to $V_{FS}/(2^n - 1)$ V. Figure 5-13 illustrates the prevailing uniform quantizing algorithm whereby an input signal that falls within the V_jth-level range of $\pm q/2$ is encoded at the V_jth level with a quantization error of ε volts. This error may range up to $\pm q/2$, and is an irreducible noise added to a converter output signal. The conventional assumption concerning the probability density function of this noise is that it is uniformly distributed along the interval $\pm q/2$, and is represented as the A/D converter quantizing uncertainty error of value 1/2 LSB proportional to converter word length.

The equivalent rms error of quantization (E_{qe}) produced by this noise is described by Equation (5-6). The rms sinusoidal signal-to-noise ratio (SNR) of Equation (5-7) then defines the output signal quality achievable, expressed in power dB, for an A/D converter of n bits with a noise-free input signal. These relationships are tabulated in Table 5-7. Equation (5-8) defines the dynamic range of a data converter of n bits in voltage dB. Converter dynamic range is useful for matching A/D converter word length in bits to a required analog input signal span to be represented digitally. For example, a 10 mV-to-10 V span (60 voltage dB) would require a minimum converter wordlength n of 10 bits. It will be shown in Section 6-3 that additional considerations are involved in the conversion of an input signal to an n-bit accuracy other than the choice of A/D converter wordlength, where the dynamic range of a digitized signal may be represented to n bits without achieving n-bit data accuracy. However, the choice of a long-wordlength A/D converter will beneficially minimize both quantization noise and A/D device error and provide increased converter linearity.

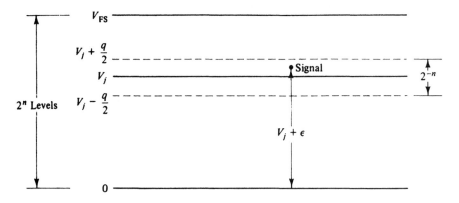

Figure 5-13. Quantization level parameters.

Table 5-7. Decimal equivalents of 2^n and 2^{-n}

Bits, n	Levels, 2^n	LSB weight, 2^{-n}	Quantization SNR, dB
1	2	0.5	8
2	4	0.25	14
3	8	0.125	20
4	16	0.0625	26
5	32	0.03125	32
6	64	0.015625	38
7	128	0.0078125	44
8	256	0.00390625	50
9	512	0.001953125	56
10	1,024	0.0009765625	62
11	2,048	0.00048828125	68
12	4,096	0.000244140625	74
13	8,192	0.0001220703125	80
14	16,384	0.00006103515625	86
15	32,768	0.000030517578125	92
16	65,536	0.0000152587890625	98
17	131,072	0.00000762939453125	104
18	262,144	0.000003814697265625	110
19	524,288	0.0000019073486328125	116
20	1,048,576	0.00000095367431640625	122

$$\text{Quantization error } E_{qe} = \left(\frac{1}{q} \int_{-q/2}^{q/2} \varepsilon^2 \cdot d\varepsilon \right) \tag{5-6}$$

$$= \frac{q}{2\sqrt{3}} \text{ rms volts}$$

$$\text{Quantization quality SNR} = 10 \log \left(\frac{V_{FS} / 2\sqrt{2}}{E_{qe}} \right)^2 \tag{5-7}$$

$$= 10 \log \left(\frac{2^n \cdot q / 2\sqrt{2}}{q / 2\sqrt{3}} \right)^2$$

$$= 6.02\, n + 1.76 \text{ power dB}$$

$$\text{Dynamic range} = 20 \log (2^n) \tag{5-8}$$

$$= 6.02 \; n \text{ voltage dB}$$

The input comparator is critical to the conversion speed and accuracy of an A/D converter, as shown in Figure 5-14. Generally, it must possess sufficient gain and bandwidth to achieve switching and settling to the amplitude error of interest ultimately determined by noise sources present, such as described in Section 4-1.

Described now are seven prevalent A/D conversion methods and their application considerations. Architectures presented include integrating dual slope, sampling successive approximation, digital angle converters, charge balancing and its evolution to oversampling sigma–delta converters, simultaneous or flash, and pipelined subranging. The performance of these conversion methods all benefit from circuit advances and monolithic technologies in their accuracy, stability, and reliability that permit expression in terms of simplified static, dynamic, and temperature-parameter error budgets, illustrated by Table 5-8.

Quantizing uncertainty constitutes converter dynamic amplitude error, illustrated by Figure 5-12(b). Mean integral nonlinearity describes the maximum deviation of the static-transfer characteristic between initial and final code transitions in Figure 5-12(a). Circuit offset, gain, and linearity temperature coefficients are combined into a single percent-of-full-scale temperature coefficient expression. Converter signal-to-noise plus distortion expresses the quality of spurious and linearity dynamic performance. This latter error is influenced by data converter -3dB frequency response, which generally must equal or exceed its conversion rate f_s to avoid amplitude and phase errors, considering the presence of input signal BW values up to the $f_s/2$ Nyquist frequency and the provisions of Tables 3-5

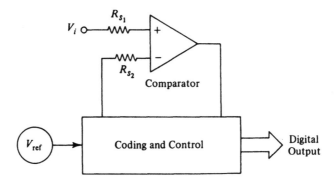

Figure 5-14. Comparator-oriented A/D converter diagram.

Table 5-8. Representative 12-bit ADC errors

Mean integral nonlinearity (1 LSB)		0.024%
Quantizing Uncertainty (1/2 LSB)		0.012%
Temperature coefficient (1 LSB)		0.024%
Noise + distortion		0.001%
$\varepsilon_{A/D}$	$\Sigma\overline{\text{mean}} + 1\ \sigma\ \text{RSS}$	0.050%

and 3-6. It is notable from Table 5-8 that the sum of the mean and RSS of converter errors provide a digital accuracy whose effective number of bits is typically one bit less than the specified converter word length.

Integrating converters provide noise rejection for the input signal at an attenuation rate of –20 dB/decade of frequency, as described in Figure 5-15, with sinc nulls at multiples of the integration period T. The ability of an integrator to pro-

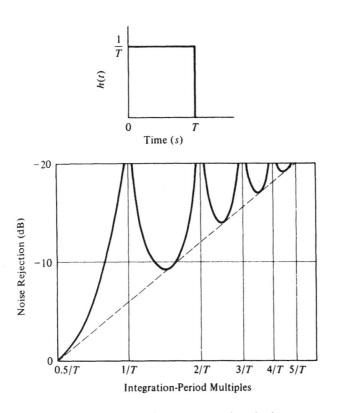

Figure 5-15. Integrating converter noise rejection.

vide this response is evident from its frequency response $H(\omega)$, obtained by the integration of its impulse response $H(t)$ in Equation (5-9). Note that this noise improvement requires integration of the signal plus noise during the conversion period and, consequently, is not furnished when a sample-hold device precedes the converter. For example, a conversion period of $16\frac{2}{3}$ ms will provide a useful null to 60-Hz interference.

$$H(\omega) = \int_o^T h(t) \cdot e^{-j\omega t} \tag{5-9}$$

$$= e^{-j\omega t} \cdot \frac{\sin \omega T/2}{\omega T/2}$$

Integrating dual-slope converters perform A/D conversion by the indirect method of converting an input signal to a representative pulse sequence that is totaled by a counter. Features of this conversion technique include self-calibration to component temperature drift, use of inexpensive components in its mechanization, and multiphasic integrations yielding improved resolution of the zero endpoint shown in Figure 5-16. Operation occurs in three steps. First is the autozero phase that stores converter analog offsets on the integrator with the input grounded. In the second, an input signal is integrated for a fixed time T_1. Finally, the input is connected to a reference of opposite polarity and integration proceeds to zero during a variable time T_2, within which clock pulses are totaled in proportion to the input signal amplitude. These operations are described by Equations (5-10) and (5-11). Integrating converters are an early method whose merits are best applied to narrow-bandwidth signals such as those encountered with hand-held multimeters. Word lengths to 16 bits are available, but conversion is limited to 1 KSPS.

$$\Delta V_1 = \frac{1}{RC} \cdot V_i \cdot T_{1_{\text{constant}}} \tag{5-10}$$

$$= \frac{1}{RC} \cdot V_{\text{ref}} \cdot T_{2_{\text{variable}}}$$

$$T_2 = \frac{V_1 \cdot T_1}{V_{\text{ref}}} \tag{5-11}$$

The successive-approximation technique is the most widely applied A/D converter type for computer interfacing, primarily because its constant conversion

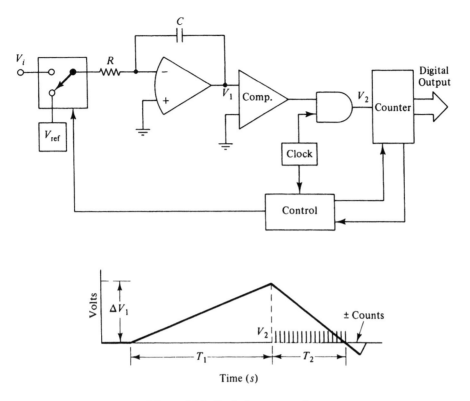

Figure 5-16. Dual-slope conversion.

period T is independent of input signal amplitude. However, it requires a preceding S/H to satisfy its requirement for a constant input signal. This feedback converter operates by comparing the output of an internal D/A converter with the input signal at a comparator, where each bit of the word length is sequentially tested during n equal time subperiods in the development of an output code representative of input signal amplitude. Converter linearity is determined by the performance of its internal D/A. Figure 5-17 describes the operation of a sampling successive-approximation converter. The conversion period and S/H acquisition time combined determine the maximum conversion rate as described in Figure 5-6. Successive-approximation converters are well suited for converting arbitrary signals, including those that are nonperiodic, in multiplexed systems. Word lengths of 16 bits are available at conversion rates to 500 KSPS.

A common method for representing angles in digital form is in natural binary weighting, where the most significant bit (MSB) represents 180 degrees and the

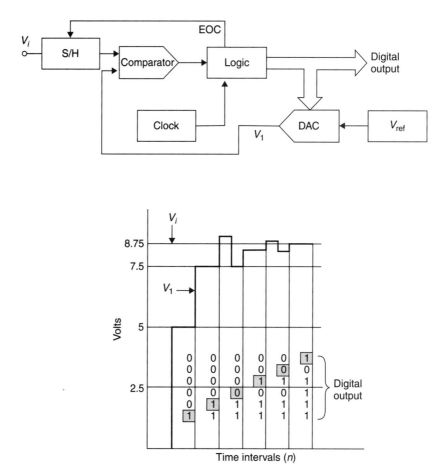

Figure 5-17. Successive-approximation conversion.

MSB-1 represents 90 degrees, as tabulated in Table 5-9. Digital synchro conversion shown in Figure 5-18 employs a Scott-T transformer connection and ac reference to develop the signals defined by Equations (5-12) and (5-13). Sine ϕ and cosine ϕ quadrature multiplications are achieved by multiplying-D/A converters whose difference is expressed by Equation (5-14). A phase-detected dc error signal, described by Equation (5-15), then pulses an up/down counter to achieve a digital output corresponding to the synchro angle θ. Related devices include digital vector generators that generate quadrature circular functions as analog outputs from digital angular inputs.

Table 5-9. Binary angle representation

Bit	Degrees
1	180
2	90
3	45
4	22.5
5	11.25
6	5.625
7	2.812
8	1.406
9	0.703
10	0.351
11	0.176
12	0.088

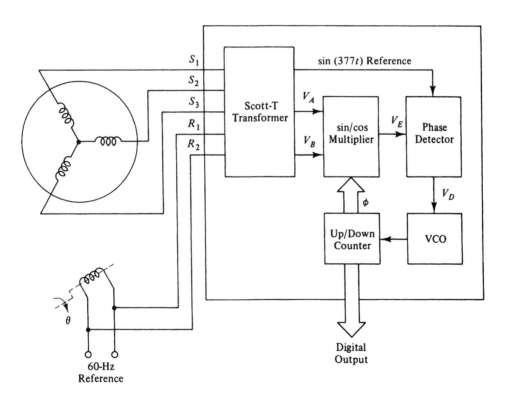

Figure 5-18. Synchro-to-digital conversion.

$$V_A = \sin(377\,t) \cdot \sin\theta \qquad\qquad (5\text{-}12)$$

$$V_B = \sin(377\,t) \cdot \cos\theta \qquad\qquad (5\text{-}13)$$

$$V_E = \sin(377\,t) \cdot \sin(\theta - \phi) \qquad\qquad (5\text{-}14)$$

$$V_D = \sin(\theta - \phi) \qquad\qquad (5\text{-}15)$$

Charge-balancing A/D converters utilize a voltage-to-frequency circuit to convert an input signal to a current I_i from which is subtracted a reference current I_{ref}. This difference current is then integrated for successive intervals, with polarity reversals determined in one direction by a threshold comparator and in the other by clock count. The conversion period for this converter is constant, but the number of count intervals per conversion vary in direct proportion to input signal amplitude, as illustrated in Figure 5-19. Although the charge-balancing converter is similar in performance to the dual-slope converter, their applications diverge; the former is compatible with and integrated in microcontroller devices.

Sigma–delta conversion employs a version of the charge-balancing converter as its first stage to perform one-bit quantization at an oversampled conversion rate f_s, whose "ones" density corresponds to analog input signal amplitude. The high quantizing noise resulting from one-bit conversion is effectively spread over a wide bandwidth from the oversampling operation which is amenable to efficient digital filtering since it is in the digital domain. The resulting spectrum is then re-sampled to an equivalent Nyquist-sampled signal bandwidth (BW) of n-bit resolution shown in Figure 5-20. Sigma–delta converters are prevalent in medium-bandwidth, high-resolution periodic signal applications from measurement instruments to telecommunications and consumer electronics. Word lengths to 20 bits for 100 kHz signal BW are available. Because of signal latency associated with oversampling and decimation operations, however, sigma–delta converters are not compatible with multiplexed applications or transient signals.

Simultaneous or flash converters are represented by the diagram of Figure 5-21 which requires $2^n - 1$ comparators biased 1 LSB apart to encode an analog input signal to n-bit resolution. All quantization levels are simultaneously compared in a single clock cycle that produces a comparator "thermometer" code with a one/zero boundary proportional to input signal amplitude. Comparator coding logic then provides a final digital output word. This architecture offers the fastest conversion rate achievable in a single clock cycle, but resolution is practically

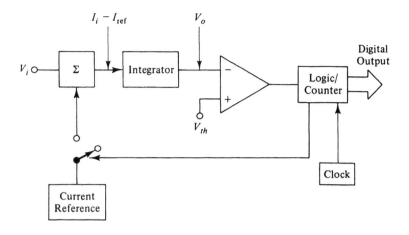

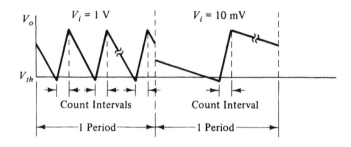

Figure 5-19. Charge-balancing conversion.

limited by the increasing number of comparators required for extending output word length. Word lengths to 10 bits with 1023 comparators are available, however, at real time rates to 100 MSPS. The flash converter beneficially can accommodate dynamic nonperiodic signals, like the slower successive approximation converter, but without an input S/H device. Applications include radar processors, electrooptical systems, and professional video.

Wideband and wide-range conversion is the province of pipelined subranging converters that offer higher resolution than flash converters with nearly the same conversion rates. Word lengths of 12 bits are common at conversion rates to 80 MSPS for applications ranging from digital spectrum analyzers to medical imaging. This architecture overcomes the comparator growth of flash converters by dividing the conversion task into cascaded stages. A typical two-stage subranging

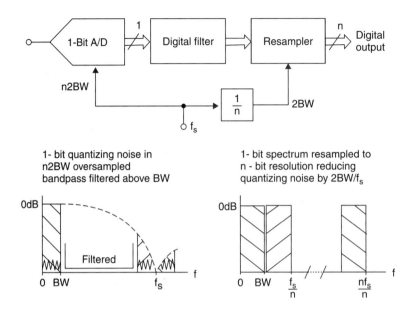

Figure 5-20. Sigma–delta conversion.

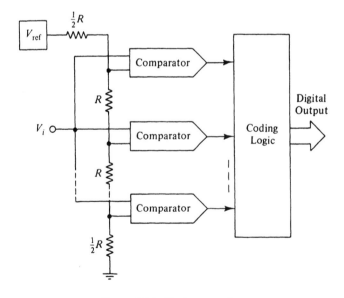

Figure 5-21. Flash conversion.

converter is shown in Figure 5-22 with two six-bit flash A/D converters requiring only 126 comparators to provide a 12-bit wordlength, where the differential subrange is converted to LSB values by the second A/D. Flash converters of m bits in p stages offer a resolution of $p \times m$ bits with $p \times (2^m - 1)$ comparators. The throughput latency of p stages of the pipeline impedes the conversion of nonperiodic signals, however.

Interrupt-initiated interfacing provides the flexibility required to accommodate asynchronous inputs from A/D converters. Upon the request of a peripheral controller, a processor interrupt is generated that initiates a service routine containing the device handler. This structure is illustrated by Figure 5-23 and offers

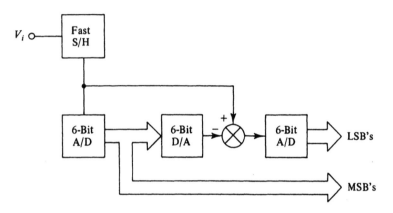

Figure 5-22. Pipelined 12-bit subranging A/D converter.

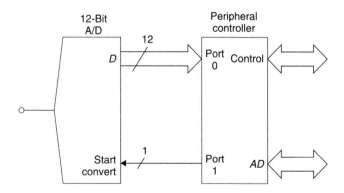

Figure 5-23. Interrupt initiated A/D interface.

enhanced throughput in comparison with program-initiated interfacing, previously described for D/A converters, by reconciling speed differences between processor buses and their peripheral controllers with FIFO buffers and DMA data transfers. Vectored multilevel-priority interrupts are also readily accommodated by this structure.

PROBLEMS

5-1. A common D/A converter employs the R-2R ladder circuit illustrated by Figure 5-9b. The utility of this network is based upon a binary division of current with excitation supplied by V_{ref}. Consider a three-bit converter with only the LSB ON and R_f value of $3R$. Sketch this circuit and determine the value of V_o, observing that a total current of $V/3R$ flows for providing resistor voltage drops.

5-2. Determine the quantization error E_{qe} in volts rms and word length n for $10\,V_{FS}$ applied to an A/D converter, sufficient to provide uniformly spaced quantizing intervals q that will accommodate a signal dynamic range of 72 dB.

5-3. Determine total device error for the 14-bit AD779k A/D converter, including an internal S/H represented by its acquisition error. Assume sufficient settling time between samples to provide a nominal S/H acquisition error of 0.01%, and evaluate error entries following the example ADC of Table 5-8. Also note the typically available binary-bit accuracy of this converter with reference to Table 6-2.

5-4. Design an eight-bit data and control bus for a digital I/O subsystem compatible with an Intel 8085 microprocessor employing discrete 8282 port devices. Diagram all connections, buses, control lines, and logic gates necessary with address lines for the input port at 40 Hex and output port at 20 Hex.

BIBLIOGRAPHY

1. Allen, R, "A/D and D/A Converters: Bridging the Analog World to the Computer," *Electronic Design News,* February 5, 1973.
2. Sheingold, D.H. (Ed.), *Analog-Digital Conversion Handbook,* Analog Devices, Norwood, Mass., 1972.
3. Artwick, B.A., *Microcomputer Interfacing,* Englewood Cliffs, NJ: Prentice-Hall, 1980.
4. Demler, M.J., *High-Speed Analog-to-Digital Conversion,* New York: Academic Press, 1991.
5. Gordon, B.M., *The Analogic Data-Conversion Systems Digest,* Analogic, Wakefield, Mass., 1977.

6. Hnatek, E.R., *A User's Handbook of D/A and A/D Converters,* New York: John Wiley, 1976.

7. Hoeschele, D.F., *Analog-to-Digital, Digital-to-Analog Conversion Techniques,* New York: Wiley, 1968.

8. Lindheimer, M., "Guidelines for Digital-to-Analog Converter Applications," *Electronic Equipment Engineering,* September 1970.

9. Stantucci, D., "Data Acquisition Can Falter Unless Components are Well Understood," *Electronics,* November 27, 1975.

10. Boyes, G.S. (Ed.), *Synchro and Resolver Conversion,* Analog Devices, Norwood, Mass., 1980.

11. Tobey, G.E., "Ease Multiplexing and A/D Conversion," *Electronic Design,* April 12, 1973.

12. Zuch, E., *Data Acquisition and Conversion Handbook,* Datel-Intersil, Mansfield, Mass., 1977.

CHAPTER 6

Sampled Data and Reconstruction with Intersample Error

6-0 INTRODUCTION

A fundamental requirement of sampled-data system is the sampling of continuous-time signals to obtain a representative set of numbers that can be used by a digital computer. The primary goal of this chapter is to provide an understanding of this process. The first section explores theoretical aspects of sampling and the formal considerations of signal recovery, including ideal Wiener filtering in signal interpolation. Aliasing of signal and noise are considered next in a detailed development involving the heterodyne basis of evaluation. This development coordinates signal bandwidth, sample rate, and band limiting prior to sampling to achieve minimum aliasing error under conditions of significant aliasable content. The third section addresses intersample error in sampled systems, and provides a sample-rate-to-signal-bandwidth ratio (f_s/BW) expressing the step-interpolator representation of sampled data in terms of equivalent binary accuracy. The final section derives a mean-squared-error criterion for evaluating the performance of practical signal recovery methods. This provides interpolated output signal accuracy in terms of the corresponding minimum required sample rate, and suggests a data conversion system design procedure based on considering system output performance requirements first.

Advanced Instrumentation and Computer I/O Design, Second Edition. By Patrick H. Garrett
Copyright © 2013 the Institute of Electrical and Electronics Engineers, Inc.

6-1   SAMPLED DATA THEORY

Observation of typical sensor signals generally reveals band-limited continuous functions with a diminished amplitude outside of a specific frequency band, except for interference or noise, which may extend over a wide bandwidth. This is attributable to the natural rolloff or inertia associated with actual processes or systems providing the sensor excitation. Sampled-data systems provide discrete signals of diminished accuracy from continuous signals of original accuracy. Of interest is how much information is lost by the sampling operation and the accuracy to which an original continuous signal can be reconstructed from its sampled values. The consideration of periodic sampling offers a mathematical solution to this problem for band-limited sampled signals of bandwidth BW. Signal discretization is illustrated for the two classifications of nonreturn-to-zero (NRZ) sampling and return-to-zero (RZ) sampling in Figure 6-1. This figure represents sampling classifications in both the time and frequency domains, where τ is the sampling function width and T the sampling period (the inverse of sample rate f_s). The determination of specific sample rates that provide sampled-data accuracies of interest is a central theme of this chapter.

The provisions of periodic sampling are based on Fourier analysis and include the existence of a minimum sample rate for which theoretically exact signal reconstruction is possible from the sampled sequence. This is significant in that signal sampling and recovery are considered simultaneously, correctly implying that the design of data conversion and recovery systems should also be considered jointly. The interpolation formula of Equation (6-1) analytically describes the approximation $\hat{x}(t)$ of a continous-time signal $x(t)$ with a finite number of samples from the sequence $x(nT)$. $\hat{x}(t)$ is obtained from the inverse Fourier transform of the input sequence, which is derived from $x(t) \cdot p(t)$ as convolved with the ideal interpolation function $H(f)$ of Figure 6-2. This results in the sinc amplitude response in the time domain owing to the rectangular characteristic of $H(f)$. Due to the orthogonal behavior of Equation (6-1), only one nonzero term is provided at each sampling instant. Contributions of samples other than ones in the immediate neighborhood of a specific sample diminish rapidly because the amplitude response of $H(f)$ tends to decrease inversely with the value of n. Consequently, the interpolation formula provides a useful relationship for describing recovered band-limited sampled-data signals, with T chosen sufficiently small to prevent signal aliasing. Aliasing is discussed in detail in the following section. Figure 6-3 shows the behavior of this interpolation formula, including its output approximation $\hat{x}(t)$.

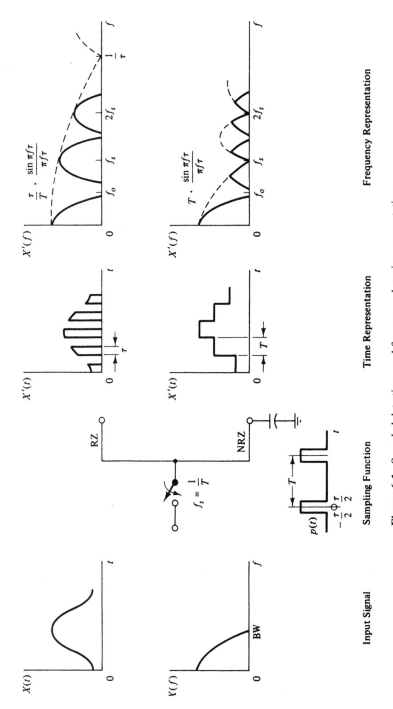

Figure 6-1. Sampled data time- and frequency-domain representation.

141

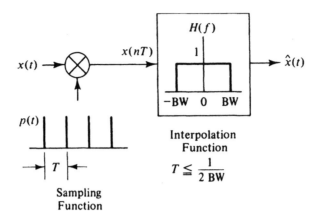

Figure 6-2. Ideal sampling and recovery.

$$\hat{x}(t) = F^{-1}\left\{f[x(nT)] * H(f)\right\} \tag{6-1}$$

$$= \sum_{n=-\infty}^{\infty} \left\{T\int_{-BW}^{BW} x(nT)\, e^{-j2\pi fT}\right\} \cdot e^{j2\pi ft} \cdot df$$

$$= T\sum_{n=-\infty}^{\infty} x(nT)\frac{e^{j2\pi BW(t-nT)} - e^{-j2\pi BW(t-nT)}}{j2\pi BW(t-nT)}$$

$$= 2TBW\sum_{n=-\infty}^{\infty} x(nT)\frac{\sin 2\pi BW(t-nT)}{2\pi BW(t-nT)}$$

A formal description of this process was provided both by Kolmogoroff and Wiener [6, 13]. It is important to note that the ideal interpolation function $H(f)$ utilizes both phase and amplitude information in reconstructing the recovered signal $\hat{x}(t)$ and is, therefore, more efficient than conventional linear filters. However, this ideal interpolation function cannot be physically realized because its impulse response $H(f)$ is noncausal, requiring an output that anticipates its input. As a result, practical interpolators for signal recovery utilize amplitude information that can be made efficient, although not optimum, by achieving appropriate weighting of the reconstructed signal. These principles are observed in Section 6-4 in the development of criteria for evaluating practical signal interpolators.

A significant consideration imposed upon the sampling operation results from the finite width τ of practical sampling functions, denoted by $p(t)$ in Figure 6-1.

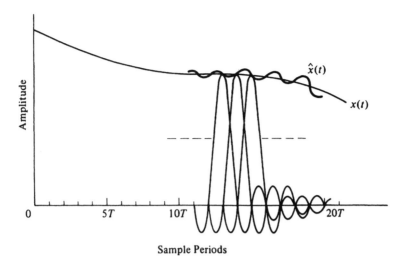

Figure 6-3. Signal interpolation.

Since the spectrum of a sampled signal consists of its original base-band spectrum $X(f)$ plus a number of images of this signal, these image signals are shifted in frequency by an amount equal to the sampling frequency f_s and its harmonics mf_s as a consequence of the periodicity of $p(t)$. The width of τ determines the amplitude of these signal images, as attenuated by the sinc functions described by the dashed lines of $X'(f)$ in Figure 6-1, for both RZ and NRZ sampling. Of particular interest is the attenuation impressed upon the baseband spectrum of $X'(f)$ corresponding to the amplitude and phase of the original signal $X(f)$. A useful criterion is to consider the average baseband amplitude error between dc and the signal BW expressed as a percentage of the full-scale departure from unity gain. Also, process bandwidth must be sufficient to support these image spectra until their amplitudes are attenuated by the sinc function to preserve signal fidelity. The mean sinc amplitude error is expressed for RZ and NRZ sampling by Equations (6-2) and (6-3). The sampled-data bandwidth requirement for NRZ sampling is generally more efficient in system bandwidth utilization than the $1/\tau$ null provided by RZ sampling. The minimization of mean sinc amplitude error may also influence the choice of f_s. The folding frequency f_0 in Figure 6-1 is an identity equal to $f_s/2$, and the specific NRZ sinc attenuation at f_0 is always 0.636, or -3.93 dB.

$$\overline{\varepsilon_{\text{RZ sinc \%FS}}} = \frac{1}{2}\left(1 - \frac{\tau}{T} \cdot \frac{\sin \pi BW\tau}{\pi BW\tau}\right) \cdot 100\% \qquad (6\text{-}2)$$

$$\overline{\varepsilon_{\text{NRZ sinc \%FS}}} = \frac{1}{2}\left(1 - \frac{\sin \pi BWT}{\pi BWT}\right) \cdot 100\% \qquad (6\text{-}3)$$

RZ sampling is primarily used for multiplexing multichannel signals into a single channel, such as encountered in telemetry systems. In Figure 6-1, the dc component of RZ sampling has an amplitude of τ/T, its average value or sampling duty cycle, which may be scaled as required by the system gain. NRZ sampling is inherent in the operation of all data-conversion components encountered in computer input–output systems, and reveals a dc component proportional to the sampling period T. In practice, this constant is normalized to unity by the $1/T$ impulse response associated with the transfer functions of actual data-conversion components.

Note that the sinc function and its attenuation with frequency in a sampled-data system is essentially determined by the duration of the sampled-signal representation $X'(t)$ at any point of observation, as illustrated in Figure 6-1. For example, an A/D converter with a conversion period T double the value employed for a following connected D/A converter will exhibit an NRZ sinc function having twice the attenuation rate versus frequency as that of the D/A, which is attributable to the transformation of the sampled-signal duration. Such D/A oversampling accordingly offers the reduced output sinc error discussed in Section 6-4. Sampled-data systems, therefore, possess a sinc function that is transformed as a consequense of changes in sampling parameters at each data conversion device in the cascade.

6-2 ALIASING OF SIGNAL AND NOISE

The effect of undersampling a continuous signal is illustrated in both the time and frequency domains in Figure 6-4. This demonstrates that the mapping of a signal to its sampled-data representation does not have an identical reverse mapping if it is reconstructed as a continuous signal when it is undersampled. Such signals appear as lower-frequency aliases of the original signal, which are defined by Equation (6-4) when $f_s < 2\,BW$. As the sample rate f_s is reduced, samples move further apart in the time domain, and signal images closer together in the frequency domain. When image spectrums overlap as illustrated in Figure 6-4b, signal aliasing occurs. The consequence of this result is the generation of intermodulation distortion that cannot be removed by later signal processing operations. Of interest is aliasing at f_0 between the baseband spectrum, representing the amplitude and

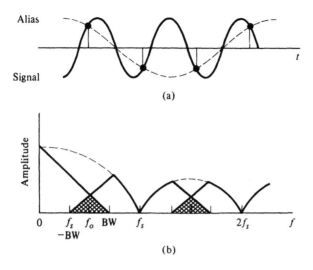

Figure 6-4. Time and frequency representation of undersampled signal aliasing.

phase of the original signal, and the first image spectrum. The folding frequency f_0 is the highest frequency at which sampled-data signals may exist without being undersampled. Accordingly, f_s must be chosen to be greater than twice the signal BW to ensure the absence of signal aliasing, which usually is readily achieved in practice.

$$f_{\text{alias}} = |f_s - BW| \qquad f_s < 2\,BW \qquad\qquad (6\text{-}4)$$
$$= \text{nonexistent} \quad f_s \geq 2\,BW$$

Of greater general concern and complexity is noise aliasing in sampled-data systems. This involves either out-of-band signal components, such as coherent interference or random noise spectra, present above f_0 and, therefore, undersampled. One or more of these sources are frequently present in most sampled-data systems. Consequently, the design of these systems should provide for the analysis of noise aliasing and the coordination of system parameters to achieve the aliasing attenuation of interest. Understanding of baseband aliasing is aided with reference to Figures 6-5 and 6-6. The noise aliasing source bands shown are heterodyned within the baseband signal between dc and f_0, derived by Equation (6-5) as $mf_s - BW \leq f_{\text{noise}} < mf_s + BW$, as a consequence of the sampling function spectra that arise at multiples of f_s. The resulting combination

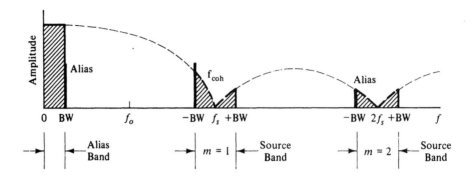

Figure 6-5. Coherent interference aliasing.

of signal and aliasing components generate intermodulation distortion proportional to the baseband alias amplitude error derived by Equations (6-7) through (6-10).

$$mf_s - BW \le f_{\text{noise}} < mf_s + BW \qquad \text{alias source frequencies} \qquad (6\text{-}5)$$

$$f_{\substack{\text{coherent} \\ \text{alias}}} = |mf_s - f_{coh}| \qquad \text{at baseband} \qquad (6\text{-}6)$$

$$= 24\ \text{Hz} - 23\ \text{Hz}$$

$$= 1\ \text{Hz}\ (m = 1)$$

$$\varepsilon_{\substack{\text{coherent} \\ \text{alias}}} = V_{\text{coh\%FS}} \cdot \text{filter attn} \cdot \text{sinc} \qquad (6\text{-}7)$$

$$= 50\%\text{FS} \cdot \frac{1}{\sqrt{1 + \left(\dfrac{f_{coh}}{f_c}\right)^{2n}}} \cdot \text{sinc}\left(\frac{|mf_s - f_{coh}|}{fs}\right)$$

$$= 50\%\text{FS} \cdot \frac{1}{\sqrt{1 + \left(\dfrac{23}{3}\right)^{6}}} \cdot \text{sinc}\left(\frac{|24 - 23|}{24}\right)$$

$$= 50\%\text{FS} \cdot (0.0024) \cdot (0.998) \tag{6-7}$$

$$= 0.12\%\text{FS with presampling filter}$$

$$N_{\text{alias}} = \sum_{0}^{\substack{\#\text{source}\\ \text{bands}}} (V_{\text{noise}}\text{rms})^2 \cdot A^2(f) \qquad \text{at baseband} \tag{6-8}$$

$$= \sum_{0}^{f_{hi}/f_s} (0.1\ V_{\text{FS}})^2 \cdot \left[\cfrac{1}{\sqrt{1 + \left(\cfrac{f_s}{f_c}\right)^{2n}}} \right]^2$$

$$= \sum_{0}^{1} (0.01\ V_{\text{FS}}^2) \cdot \left[\cfrac{1}{\sqrt{1 + \left(\cfrac{24}{3}\right)^{6}}} \right]^2$$

$$= 0.038 \times 10^{-6} \cdot V_{\text{FS}}^2 \text{ watt into } 1\ \Omega$$

$$\text{SNR}_{\substack{\text{random}\\ \text{alias}}} = \frac{V_s^2\ \text{rms}}{N_{\text{alias}}} \tag{6-9}$$

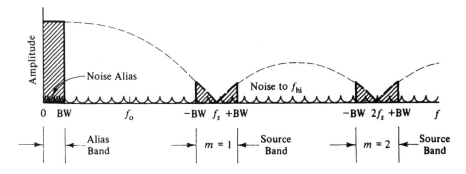

Figure 6-6. Random interference aliasing.

$$\varepsilon_{\substack{\text{random} \\ \text{alias}}} = \frac{\sqrt{2} \cdot 100\%}{\sqrt{\text{SNR}_{\substack{\text{random} \\ \text{alias}}}}} \tag{6-10}$$

$$= \frac{\sqrt{2} \cdot 100\%}{\sqrt{V_{FS}^2 / 0.038 \times 10^{-6} V_{FS}^2}}$$

$$= 0.027\%\text{FS with presampling filter}$$

Coherent alias frequencies capable of interfering with baseband signals are defined by Equation (6-6). The amplitude of the aliasing error components expressed as a percent of full scale are provided for both NRZ and RZ sampling by Equation (6-7) with the appropriate sinc function argument. Note that this equation may be evaluated to determine the aliasing amplitude error with or without presampling filtering and its effect on aliasing attenuation. For example, consider a 1-Hz signal BW for a NRZ sampled-data system with an f_s of 24 Hz. A 23-Hz coherent interfering input signal of –6 dB amplitude (50%FS) will be heterodyned both to 1 Hz and 47 Hz by this 24-Hz sampling frequency, with negligible sinc attenuation at 1 Hz and approximately –30 dB at 47 Hz, for a coherent aliasing base-band aliasing error of 50%FS applying Equation (6-7) in the absence of a presampling filter. This is illustrated by Figure 6-5. The addition of a lowpass three-pole Butterworth presampling filter with a 3-Hz cutoff frequency, to minimize filter error to $\overline{0.1\%\text{FS}}$ over the signal BW, then provides –52 dB attenuation to the 23-Hz interfering signal for a negligible 0.12%FS base-band aliasing error shown by the calculations accompanying Equation (6-7). This filter may be visualized superimposed on that figure.

A more complex situation is presented in the case of random noise because of its wideband spectral characteristic. This type of interference exhibits a uniform amplitude representing a Gaussian probability distribution. Aliased base-band noise power N_{alias} is determined as the sum of heterodyned noise source bands between $mf_s - \text{BW} \le f_{\text{noise}} < mf_s + \text{BW}$. These bands occur at intervals of f_s in frequency, shown in Figure 6-6, up to a –3 dB band-limiting f_{hi}, such as provided by an input amplifier cutoff frequency preceding the sampler, with f_{hi}/f_s total noise source bands contributing. N_{alias} may be evaluated with or without the attenuation provided by a presampling filter $A(f)$ in determining baseband random noise aliasing error, which is expressed as an aliasing signal-to-noise ratio in Equations (6-9) and (6-10). The small sinc amplitude attenuation at base band is omitted for simplicity.

Consider a –20 dB (0.1 FS) example V_{noise} rms level extending from dc to an f_{hi} of 1 kHz. Solution of Equations (6-8) through (6-10), in the absence of a filter with $A(f) = 1.0$, yields 0.42 V_{FS}^2 watt into 1 ohm as N_{alias} with an f_s of 24 Hz and 42 source bands contributing up to 1 kHz for a random noise aliasing error of 90%FS. Consideration of the previous 1-Hz signal BW and 3-Hz cutoff, three-pole Butterworth low-pass filter provides –54 dB average attenuation over the first noise source band centered at f_s. Significantly greater filter attenuation is imposed at higher noise frequencies, resulting in negligible contribution from those source bands to N_{alias}. The presampling filter effectiveness, therefore, is such that the random noise aliasing error is only 0.027%FS. It is notable that base-band aliased noise inversely follows the response of $A(f)$ to f_0, owing to heterodyned noise contributions from f_0 to f_s.

Table 6-1 offers an efficient coordination of presampling filter specifications employing a conservative criterion of achieving –40 dB input attenuation at f_0 in terms of a required f_s/BW ratio that defines the minimum sample rate for preventing noise aliasing. The foregoing coherent and random noise aliasing examples meet these requirements with their f_s/BW ratios of 24, employing the general-application three-pole Butterworth presampling filter, whose cutoff frequency f_c of three times signal BW provides only a nominal device error addition while achieving significant antialiasing protection. RC presampling filters are clearly the least efficient and appropriate only for dc signals, considering their required f_s/BW ratio to obtain useful aliasing attenuation. Six-pole Butterworth presampling filters are most efficient in conserving sample rate while providing equal aliasing attenuation at the cost of greater filter complexity. A three-pole Bessel filter is unparalleled in its linearity to both amplitude and phase for all signal

Table 6-1. Coordination of sample rate, signal bandwidth, and sinc function with presampling filter for aliasing attenuation at the folding frequency

Application	Presampling filter poles			f_s/BW for –40 dB attenuation at f_0 including –4 dB sinc and filter f_c of			Filter $\overline{\varepsilon_{\%FS}}$ per signal type	
	RC	Bessel	Butterworth	20 BW	10 BW	3 BW	DC, Sines	Harmonic
DC signals	1			2560			0.10	1.20
Linear phase		3			80		0.10	0.10
General			3			24	0.10	0.11
Brickwall			6			12	0.05	0.15

types as an antialiasing filter, but requires an inefficient f_s/BW ratio of 80 to compensate for its passband amplitude rolloff. The following sections consider the effect of sample rate on sampled-data accuracy, first as step-interpolated data principally encountered on a computer data bus, and then including post-D/A converter interpolation-function-assisted signal reconstruction.

6-3   SAMPLED DATA INTERSAMPLE AND APERTURE ERRORS

The NRZ-sampling step-interpolated data representation of Figure 6-7 denotes the way converted data are handled in digital computers, whereby the present sample is current data until a new sample is acquired. Both intersample and aperture volts ΔV_{pp} and $\Delta V'_{pp}$, respectively, are derived in this development as time-amplitude relationships to augment that understanding.

In real-time data conversion systems, the sampling process is followed by quantization and encoding, all of which are embodied in the A/D conversion process described by Figure 5-11. Quantization is a measure of the number of discrete amplitude levels which may be assigned to represent a signal waveform, and is proportional to A/D converter output word length in bits. A/D quantization levels are uniformly spaced between 0 and V_{FS} with each being equal to the LSB interval as described in Figure 5-12. For example, a 12-bit A/D converter provides a quantization interval proportional to 0.024%FS. This example converter wordlength thus provides quantization which is sufficiently small to permit intersample error to be evaluated independently without the influence of sampling effects even though the source of both is the A/D converter. Note that both intersam-

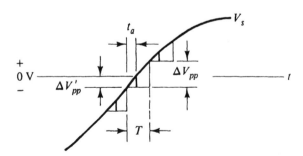

Figure 6-7. Intersample and aperture error representation.

ple and aperture error are system errors, whereas quantization error is a part of the A/D converter device error.

NRZ sampling is inherent in the operation of S/H, A/D, and D/A devices by virtue of their step-interpolator sampled data representation. Equation (6-11) describes the impulse response for this data representation in the derivation of a frequency-domain expression for step-interpolator amplitude and phase. Evaluation of the phase term at the sample rate f_s discloses that an NRZ-sampled signal exhibits an average time delay equal to $T/2$ with reference to its input. This linear phase characteristic is illustrated in Figure 6-8. The sampled input signal is acquired as shown in Figure 6-9a, and represented as discrete values in a digitally encoded form. Figure 6-9b describes the average signal delay with reference to its input of Figure 6-9a. The difference between this average signal delay and its step-interpolator representation constitute the peak-to-peak intersample error constructed in Figure 6-9c.

$$g(t) = U(t) - U(t - T) \qquad (6\text{-}11)$$

$$g(s) = \frac{1}{s} - \frac{e^{-T}}{s}$$

$$g(j\omega) = \frac{1 - e^{-j\omega T}}{j\omega}$$

$$= T\frac{\sin \pi fT}{\pi fT} \; \underline{/{-j\omega\,T/2}} \quad \text{NRZ impulse response}$$

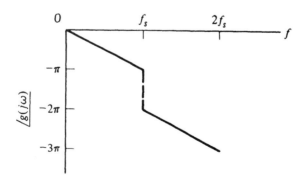

Figure 6-8. Step-interpolator phase.

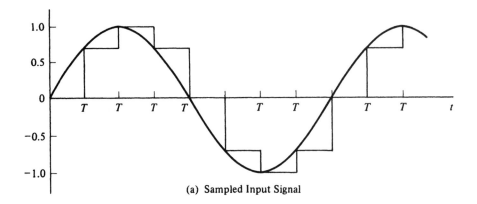

(a) Sampled Input Signal

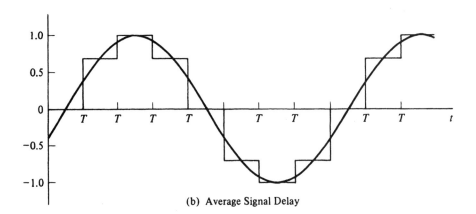

(b) Average Signal Delay

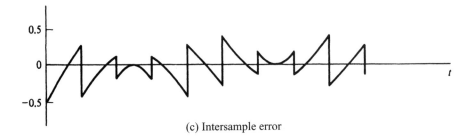

(c) Intersample error

Figure 6-9. Step-interpolator signal representation.

Evaluating delay at $f_s = \dfrac{1}{T}$:

$$\underline{/g(j\omega)} = -\pi$$

$$= 2\pi \, \text{f} t$$

$$\therefore t = -\frac{T}{2} \, \text{sec} \qquad \text{sampled signal delay}$$

Equation (6-12) describes the intersample volts ΔV_{pp} for a peak sinusoidal signal V_s evaluated at its maximum rate-of-change zero crossing shown in Figure 6-7. This representation is converted to ΔV_{rms} through normalization by $2\sqrt{5}$ from the product of the $2\sqrt{2}$ sinusoidal pp-rms factor and the $\sqrt{2.5}$ crest factor triangular step-interpolation contribution of Figure 6-9c. This expression is also equal to the square root of mean-squared error, which is minimized as the true signal and its sampled data representation converge. Equation (6-13) reexpresses Equation (6-12) to define a more useful amplitude error $\varepsilon_{\Delta V\%FS}$ represented in terms of binary equivalent values in Table 6-2, and is then rearranged in terms of a convenient f_s/BW ratio for application purposes. Describing the signal V_s relative to the specific V_{FS} scaling also permits accommodation of the influence of signal amplitude on the representative rms intersample error of a digitized waveform. Intersample error thus represents the departure of A/D output data from their corresponding continuous input signal values as a consequence of converter sampling, quantizing, and encoding functions, and signal bandwidth and amplitude dynamics.

$$\Delta V_{pp} = T \cdot \frac{dV_s}{dt} \quad \text{intersample volts} \qquad (6\text{-}12)$$

$$= T \cdot \frac{d}{dt} \, V_s \, \sin 2\pi \, BWt \Big|_{t=0}$$

$$= 2\pi \, T \, BW \, V_{spk}$$

$$\Delta V_{\text{rms}} = \frac{2\pi \, T \, BW \, V_{spk}}{2\sqrt{5}}$$

$$= \sqrt{\text{MSE}} \, \text{volts}$$

Table 6-2. Step-interpolated sampled data equivalents

Binary bits (accuracy)	Intersample error $\varepsilon_{\Delta V\%FS}$ (1LSB)	f_s/BW (numerical)	Applications
0	100.0	2	Nyquist limit
1	50.0		
2	25.0		
3	12.5		
4	6.25	32	Digital toys
5	3.12		
6	1.56		
7	0.78		
8	0.39	512	Video systems
9	0.19		
10	0.097		
11	0.049		
12	0.024	8192	Industrial I/O
13	0.012		
14	0.006		
15	0.003		
16	0.0015	131, 072	Instrumentation
17	0.0008		
18	0.0004		
19	0.0002		
20	0.0001	2, 097, 152	High-end audio

$$\varepsilon_{\Delta V\%FS} = \frac{\Delta V_{rms}}{V_{FS_{pk}} / \sqrt{2}} \cdot 100\% \quad \text{intersample error} \qquad (6\text{-}13)$$

$$= \frac{\sqrt{2}\,\pi\,BW\,V_{S_{pk}}}{\sqrt{5}\,f_s\,V_{FS_{pk}}} \cdot 100\%$$

$$\frac{f_s}{BW} = \frac{\sqrt{2}\pi\,V_{S_{pk}}\,100\%}{\sqrt{5}\,\varepsilon_{\Delta V\%FS}\,V_{FS_{pk}}} \quad \text{step-interpolated data}$$

Determining the step-interpolated intersample error of interest is aided by Table 6-2 and Equation (6-13). For example, eight-bit binary accuracy requires an f_s/BW ratio of 512 considering its LSB amplitude value of 0.39%FS. This implies sampling a sinusoid uniformly every 0.77 degree with the waveform peak amplitude scaled to the full-scale value. This obviously has an influence on the design of sampled-data systems and the allocation of their resources to achieve an intersample error of interest. With harmonic signals, the tenth-harmonic amplitude value typically declines to one-tenth that of the fundamental frequency amplitude such that intersample error remains constant between these signal frequencies for arbitrary sample rates. The f_s/BW ratio of two provides an intersample error reference defining frequency sampling that is capable of quantifying only signal polarity changes. Unlike digital measurement and control systems in which quantitative amplitude accuracy is of interest, frequency sampling is employed for information that is encoded in terms of signal frequencies encountered in communications systems that usually involve qualitative interpretation. For example, digital telephone systems often employ seven-bit accuracy to meet a human sensory error/distortion perception threshold generally taken as 0.7%FS, whose signal reconstruction accuracy is increased over that of step-interpolated data by postfilter interpolation (introduced in Section 6-4).

Figure 6-10 describes an elementary digital error budget example of 11-bit binary accuracy that is ample for a three-decimal-digit dc digital voltmeter whose step-interpolated 3.32 bits/digit requires only 10 bits for display. This acquisition system can accommodate a signal bandwidth to 10 mHz at a sample rate of 60 Hz with an f_s/BW of 6000. From Chapter 5, intrinsic noise rejection of the integrating A/D converter beneficially provides amplitude nulls to possible voltmeter interference at the f_s value of 60 Hz, and -20 dB/decade rolloff to other input frequencies.

Aperture time t_a describes the finite amplitude uncertainty $\Delta V'_{pp}$ within which a sampled signal is acquired such as by a S/H device, referencing Figure 6-7 and Equation (6-14), that involves the same relationships expressed in Equation (6-12). Otherwise, sampling must be accomplished by a device whose performance is not affected by input signal change during acquisition such as an integrating A/D. In that direct conversion case, t_a is identical to the sampling period T. A principal consequence of aperture time is the superposition of an additional sinc function on the sampled-data spectrum. The mean aperture error over the base-band signal described by Equation (6-15), however, is independent of the mean sinc error defined by Equation (6-3). Although intersample and aperture performance are similar in their relationships, variation in t_a has no influence on intersample error. For example, a fast S/H preceding an A/D converter can provide a small aperture uncertainty, but intersample error continues to be determined by the sampling period T. Figure 6-11 is a nomograph of Equation (6-14) that describes aperture error in terms of binary accuracy. Aperture error is negligible in most data conversion systems and consequently not included in the error summary.

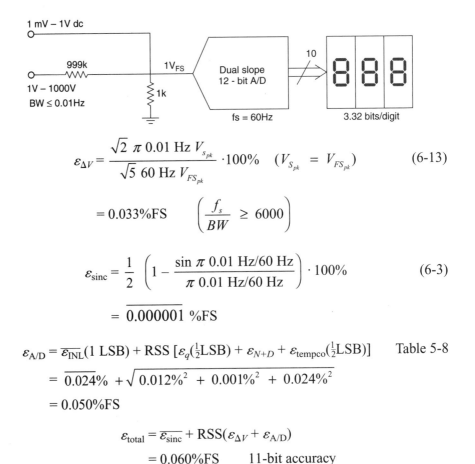

$$\varepsilon_{\Delta V} = \frac{\sqrt{2}\ \pi\ 0.01\ \text{Hz}\ V_{S_{pk}}}{\sqrt{5}\ 60\ \text{Hz}\ V_{FS_{pk}}} \cdot 100\% \quad (V_{S_{pk}} = V_{FS_{pk}}) \tag{6-13}$$

$$= 0.033\%\text{FS} \qquad \left(\frac{f_s}{BW} \geq 6000\right)$$

$$\varepsilon_{\text{sinc}} = \frac{1}{2}\ \left(1 - \frac{\sin \pi\ 0.01\ \text{Hz}/60\ \text{Hz}}{\pi\ 0.01\ \text{Hz}/60\ \text{Hz}}\right) \cdot 100\% \tag{6-3}$$

$$= \overline{0.000001}\ \%\text{FS}$$

$$\varepsilon_{\text{A/D}} = \overline{\varepsilon_{\text{INL}}}(1\ \text{LSB}) + \text{RSS}\ [\varepsilon_q(\tfrac{1}{2}\text{LSB}) + \varepsilon_{N+D} + \varepsilon_{\text{tempco}}(\tfrac{1}{2}\text{LSB})] \qquad \text{Table 5-8}$$

$$= \overline{0.024}\% + \sqrt{0.012\%^2 + 0.001\%^2 + 0.024\%^2}$$

$$= 0.050\%\text{FS}$$

$$\varepsilon_{\text{total}} = \overline{\varepsilon_{\text{sinc}}} + \text{RSS}(\varepsilon_{\Delta V} + \varepsilon_{\text{A/D}})$$

$$= 0.060\%\text{FS} \qquad 11\text{-bit accuracy}$$

Figure 6-10. Digital dc voltmeter error budget.

$$\Delta V'_{pp} = 2\pi t_a\ BWV_{spk} \quad \text{aperture volts} \tag{6-14}$$

$$\overline{\varepsilon_{a\%\text{FS}}} = \frac{1}{2}\ \left(1 - \frac{\sin \pi\ BW\ t_a}{\pi\ BW\ t_a}\right) \cdot 100\% \tag{6-15}$$

6-4   OUTPUT SIGNAL INTERPOLATION FUNCTIONS

The recovery of continuous analog signals from discrete digital signals is required in the majority of instrumentation applications. Signal reconstruction

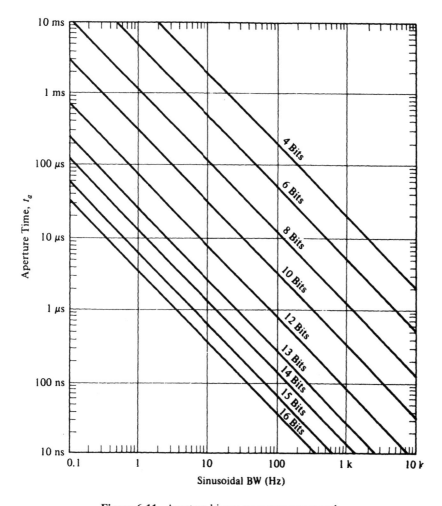

Figure 6-11. Aperture binary accuracy nomograph.

may be viewed from either time-domain or frequency-domain perspectives. In time-domain terms, recovery is similar to interpolation techniques in numerical analysis involving the generation of a locus that reconstructs a signal by connecting discrete data samples. In the frequency domain, efficient signal recovery involves band-limiting a D/A output with a low-pass postfilter to attenuate image spectra present above the baseband signal. It is of further interest to pursue signal reconstruction methods that are more efficient in sample rate requirements than the step-interpolator signal representation described by Table 6-2.

Figure 6-12 illustrates direct-D/A signal recovery with extensions that add both linear interpolator and postfilter functions. Signal delay is problematic in digital control systems, so a direct-D/A output is employed with image spectra attenuation achieved by the associated process closed-loop bandwidth. Linear interpolation is a capable reconstruction function, but achieving a nominal device error is problematical. Linear interpolator effectiveness is defined by first-order polynomials whose line-segment slopes describe the difference between consecutive data samples. A pre-D/A converter software methodology is also described for output interpolation by Figure 6-16.

Figure 6-13 shows a frequency-domain representation of a sampled signal of bandwidth BW with images about the sampling frequency f_s. This ensemble illustrates image spectra attenuated by the sinc function and low-pass postfilter in achieving convergence of the total sampled data ensemble to its ideal base band BW value. An infinite-series expression of the image spectra summation is given by Equation (6-16) that equals the mean squared error (MSE) for direct-D/A output. It follows that step-interpolated signal intersample error may be evaluated by Equation (6-17), employing this MSE in deriving the first output interpolator function of Table 6-3, whose result corresponds identically to that of Equations (6-12) and (6-13). Note that the sinc terms of Equation (6-17) are evaluated at the worst-case first-image maximum-amplitude frequencies of $f_s \pm BW$.

$$\text{MSE} = V_s^2 \sum_{k=1}^{\infty} \left[\text{sinc}^2 \left(k - \frac{BW}{f_s} \right) + \text{sinc}^2 \, k \left(1 + \frac{BW}{f_s} \right) \right] \quad \text{D/A output} \quad (6\text{-}16)$$

$$= 2 \, V_s^2 \left[\text{sinc}^2 \left(1 - \frac{BW}{f_s} \right) + \text{sinc}^2 \left(1 + \frac{BW}{f_s} \right) \right]$$

$$\varepsilon_{\Delta V\%\text{FS}} = \left[\frac{V_{o_{FS}}^2}{2 \, V_s^2 \left[\text{sinc}^2 \left(1 - \frac{BW}{f_s} \right) + \text{sinc}^2 \left(1 + \frac{BW}{f_s} \right) \right]} \right]^{-1/2} \cdot 100\% \quad (6\text{-}17)$$

The choice of interpolator function should include a comparison of realizable signal intersample error and the error addition provided by the interpolator device with the goal of realizing not greater than parity in these values. Figure 6-14 shows a comparison of four output interpolators for an example sinusoidal signal at a modest f_s/BW ratio of 10. The three-pole Butterworth postfilter is especially versatile for image spectra attenuation with dc, sinusoidal, and harmonic signals and

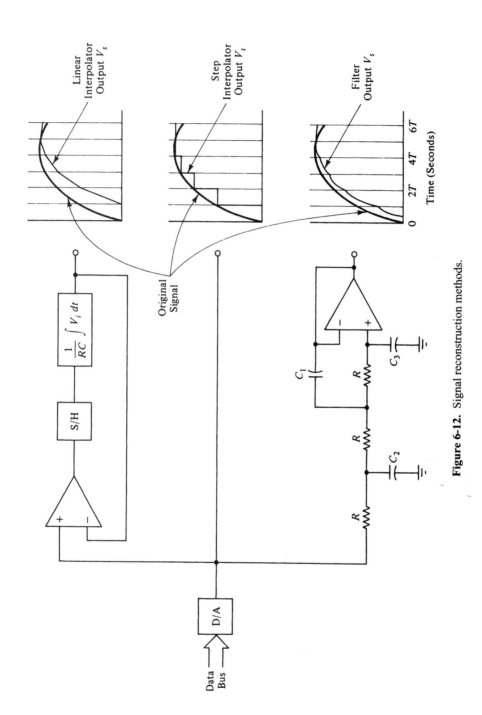

Figure 6-12. Signal reconstruction methods.

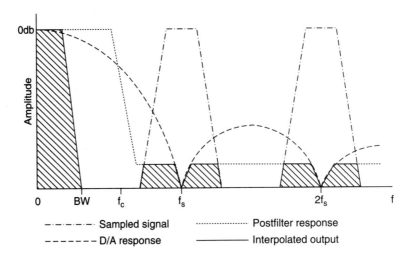

Figure 6-13. Signal recovery spectral ensemble.

Table 6-3. Output interpolation functions

Interpolator	Amplitude	Intersample error, $\varepsilon_{\Delta V}$%FS
D/A	$\mathrm{sinc}\,(f/f_s)$	$\left[\dfrac{V_{0_{FS}}^2}{2\,V_s^2\left[\mathrm{sinc}^2\!\left(1-\dfrac{BW}{f_s}\right)+\mathrm{sinc}^2\!\left(1+\dfrac{BW}{f_s}\right)\right]}\right]^{-1/2}\cdot 100\%$
D/A + linear	$\mathrm{sinc}^2(f/f_s)$	$\left[\dfrac{V_{0_{FS}}^2}{V_s^2\left[\mathrm{sinc}^4\!\left(1-\dfrac{BW}{f_s}\right)+\mathrm{sinc}^4\!\left(1+\dfrac{BW}{f_s}\right)\right]}\right]^{-1/2}\cdot 100\%$
D/A + 1-pole RC	$\mathrm{sinc}(f/f_s)[1+(f/f_c)^2]^{-1/2}$	
D/A + Butterworth n-pole low-pass	$\mathrm{sinc}(f/f_s)[1+(f/f_c)^{2n}]^{-1/2}$	$\left[\dfrac{V_{0_{FS}}^2}{V_s^2\left[\mathrm{sinc}^2\!\left(1-\dfrac{BW}{f_s}\right)\left[1+\left(\dfrac{f_s-BW}{f_c}\right)^{2n}\right]^{-1}+\mathrm{sinc}^2\!\left(1+\dfrac{BW}{f_s}\right)\left[1+\left(\dfrac{f_s+BW}{f_c}\right)^{2n}\right]^{-1}\right]}\right]^{-1/2}\cdot 100\%$

$$f_s \pm BW \text{ substituted for } f$$

adds only nominal device error with reference to Tables 3-5 and 3-6. Its six-bit improvement over direct-D/A recovery is substantial with significant convergence toward ideal signal reconstruction. Bessel filters require an f_c/BW of 20 to obtain a nominal device error, which limits their effectiveness to that of a one-pole RC. However, this does not diminish the utility of Bessel filters for conditioning phase signals. Interpolator residual intersample error values revealed at fractional BW in Figure 6-14 also define the minimum data wordlengths necessary to preserve the interpolated signal accuracy achieved. Table 6-4 describes the signal time delay encountered in transit through the respective interpolation functions.

Oversampled data conversion, introduced by the sigma–delta A/D converter of Figure 5-20, relies upon the increased quantization SNR of 6 dB for each fourfold increase in f_s enabling one binary-bit-equivalent of additional SNR performance from Table 5-7. The merit of oversampled D/A conversion, compared to Nyquist sampling and recovery, where signal BW may exist up to the folding frequency value of $f_s/2$, is a comparable output SNR improvement without increasing the converter word length, accompanied by reduced sinc attenuation with reference to Equation (6-3). Figure 6-15 shows the performance improvement of four times oversampling D/A conversion with the sampled signal present every fourth sample. First, the fixed quantization noise power for any D/A word length is now distributed over four times the spectral occupancy such that only one-

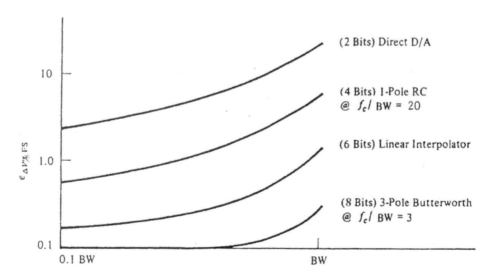

Figure 6-14. Output interpolator comparison (sinusoid, $f_s/BW = 10$).

Table 6-4. Interpolation transfer delay

Interpolator	Time (seconds)
D/A	$\dfrac{1}{2f_s}$
D/A + linear	$\dfrac{1}{2f_s} + \dfrac{1}{f_s}$
D/A + Butterworth n-pole	$\dfrac{1}{2f_s} + \dfrac{n}{4f_c}$

fourth of this noise is in the signal bandpass to BW yielding a 6 dB SNR improvement. Further, the accompanying sinc amplitude attenuation at signal BW is –0.22 dB compared to –3.9 dB encountered with Nyquist sampled signals. Equivalent oversampling performance additions may be obtained at 16 f_s, 64 f_s, and higher multiples. However, while oversampling achieves reduced quantization noise and sinc attenuation it cannot increase signal data content.

Pre-D/A converter software processing for output signal interpolation has experimentally demonstrated two binary bits of data accuracy enhancement compared

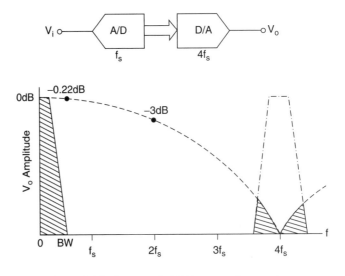

Figure 6-15. Oversampled D/A conversion spectrum.

with direct-D/A step-interpolation for essentially identical time delay, as shown by Hoffman [4]. This is especially of value in control-loop applications, where increased signal delay of post-D/A interpolation filters would degrade control stability. The illustrated binary accumulation interpolating D/A converter of Figure 6-16 reveals an interstep oscillation pattern of signal energy and frequency content resulting from a sinusoidal excitation waveform. The four MSB values of this figure provide a frequency of 2^4 times the input data sample rate f_s in register R_3. When this sum exceeds $2^4 - 1$, a carry is generated and summed with the LSB values in register R_4, where the eight-bit D/A output is updated at $16 f_s$. Intel 8085 microcomputer assembler code is employed for this prototype, as shown in Table 6-5.

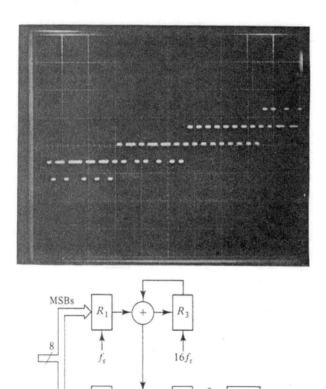

Figure 6-16. Binary accumulation pre-D/A interpolation.

Table 6-5. Binary accumulation assembler code

```
        SUB     A           ; SET A TO 0
        OUT     20H         ; MAKE PORT 21H ALL INPUTS
        STA     20FFH
        MVI     A,0FFH      ; MAKE PORT 1 ALL OUTPUTS
        OUT     03H
        ANI     04H
        OUT     02H         ; MAKE PORT 0 ALL INPUTS EXCEPT BIT 3
                00H         ; SET LOAD LINE HIGH

RESET:
        MVI     E, 08H      ; INITIALIZE ADDEND TO 8 (2EXP (M-1))
        MVI     H, 10H      ; INITIALIZE OUTPUT COUNTER (2EXP (N))
INPUT:
        IN      0
        ANI     01H
        JZ      INPUT       ;WAIT FOR A/D READY
        IN      21H         ; INPUT 8 BIT WORD FROM A/D
        MOV     D, A        ; SAVE IN D
        ANI     0F0H        ; MASK OUT THE 4 LSB'S
        MOV     B, A        ; SAVE MSB'S IN B
        MOV     A, D        ; GET INPUT 8 BITS AGAIN
        ANI     0FH         ; MASK OUT THE 4 MSB'S
        MOV     C, A        ; SAVE LSB'S IN C
ADDPR:
        MOV     A, E        ; GET THE ADDEND
        ADD     C           ; ADD THE LSB'S
        MOV     D, A        ; SAVE THE ADDITION INCLUDING POSSIBLE CARRY
        ANI     0FH         ; CONVERT ADDITION MODULO 2EXP(N) = 16
        MOV     E, A        ; SAVE NEW ADDEND
        MOV     A, B        ; GET MSB'S
        ADD     D           ; ADD SUM OF ADDEND AND LSB'S WITH POSSIBLE CY
        ANI     0F0H        ; FORCE LSB'S TO 0
        JNZ     NOVER       ; IF NZ, THERE WAS NO OVERFLOW
        MOV     A, B        ; WAS AN OVERFLOW, RESTORE ORIGINAL MSB'S
NOVER:
        MOV     D, A        ; SAVE DATA TO OUTPUT, MSB'S = DATA, LSB'S = 0
OUTOK:
        IN      0H          ; TEST FOR OUTOK PULSE
        ANI     2H
        JZ      OUTOK
        MOV     A, D        ; GET DATA TO OUTPUT
        OUT     1H
        SUB     A           ; SET A TO 0
        OUT     0           ; SEND LOAD LINE LOW
        ORI     4H          ; SEND LOAD LINE HIGH
        OUT     0
        DCR     H           ; DECREMENT MODULO 16 LOOP COUNTER
        JNZ     ADDPR       ; DO 16 TIMES
        JMP     RESET       ; ELSE GET NEXT INPUT DATA

        END
```

6-5   VIDEO SAMPLING AND RECONSTRUCTION

Industrial machine vision, laboratory spectral analysis, and medical imaging instrumentation are all supported by advances in digital signal processing that are frequently coupled to television standards and computer graphics technology. Real-time imaging systems usefully employ line-scanned television standards such as RS-343A and RS-170 that generate 30 frames per second, with 525 lines per frame interlaced into one even-line and one odd-line field per frame. Each line has a sweep rate of 53.3 μsec, plus 10.2 μsec for horizontal retrace. The bandwidth required to represent discrete picture elements (pixels) is therefore determined by the discrimination of active and inactive pixels of equal width in time along a scanning line.

The implementation of a high-speed data conversion system is largely a wideband analog design task. Baseline considerations include employing data converters possessing intrinsic speed with low spurious performance. The example ADS822 A/D converter by Burr-Brown is capable of a 40 megasample-per-second conversion rate employing a pipelined architecture for input signals up to 10-MHz bandwidth, with a 10-bit output word length that limits quantization noise to −60 dB. A one-pole RC input filter with a 15-MHz cutoff frequency is coincident with the conversion-rate folding frequency f_0 to provide antialiasing attenuation of wideband input noise.

Figure 6-17 reveals that the performance of this video imaging system is dominated by intersample error that achieves a five-bit binary accuracy, providing 32 luminance levels for each reconstructed pixel. A detailed system-error budget, therefore, will not reveal additional influence on this performance. An Analog

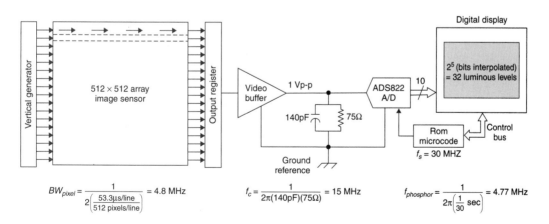

$$BW_{pixel} = \frac{1}{2\left(\dfrac{53.3 \mu s/line}{512\ pixels/line}\right)} = 4.8\ MHz \qquad f_c = \frac{1}{2\pi(140pF)(75\Omega)} = 15\ MHz \qquad f_{phosphor} = \frac{1}{2\pi\left(\dfrac{1}{30}\ sec\right)} = 4.77\ MHz$$

Figure 6-17. Video conversion and reconstruction.

Devices 10-bit ADV7128 pipelined D/A converter with a high-impedance video current output is a compatible data reconstructor providing glitchless performance. Video interpolation is achieved by the time constant of the video display for image reconstruction, whose performance is comparable to the response of a single-pole low-pass filter constrained by the 30-frames-per-second television standard.

$$\varepsilon_{\Delta V} = \left[\cfrac{V_{0_{FS}}^2}{\left\{ V_S^2 \cdot \left[\mathrm{sinc}^2 \left(1 - \dfrac{BW_{\mathrm{pixel}}}{f_s} \right) \cdot \left[1 + \left(\dfrac{f_s - BW_{\mathrm{pixel}}}{f_{\mathrm{phosphor}}} \right)^2 \right]^{-1} \right] \right. }{ \left. + \mathrm{sinc}^2 \left(1 + \dfrac{BW_{\mathrm{pixel}}}{f_s} \right) \left[1 + \left(\dfrac{f_s + BW_{\mathrm{pixel}}}{f_{\mathrm{phosphor}}} \right)^2 \right]^{-1} \right\} } \right]^{-1/2} \cdot 100\% \qquad \text{(Table 6-3)}$$

$$= \left[\cfrac{1\,V^2}{\left\{ 1\,V^2 \cdot \left[\left[\dfrac{\sin \pi \left(1 - \dfrac{4.8\,\mathrm{M}}{30\,\mathrm{M}} \right)}{\pi \left(1 - \dfrac{4.8\,\mathrm{M}}{30\,\mathrm{M}} \right)} \right]^2 \cdot \left[1 + \left(\dfrac{30\,\mathrm{M} - 4.8\,\mathrm{M}}{4.77\,\mathrm{M}} \right)^2 \right]^{-1} \right] \right. }{ \left. + \left[\dfrac{\sin \pi \left(1 + \dfrac{4.8\,\mathrm{M}}{30\,\mathrm{M}} \right)}{\pi \left(1 + \dfrac{4.8\,\mathrm{M}}{30\,\mathrm{M}} \right)} \right]^2 \cdot \left[1 + \left(\dfrac{30\,\mathrm{M} - 4.8\,\mathrm{M}}{4.77\,\mathrm{M}} \right)^2 \right]^{-1} \right\} } \right]^{-1/2} \cdot 100\%$$

$$= \left[\cfrac{1}{\left(\dfrac{0.482}{2.636} \right)^2 \cdot (0.034) + \left(\dfrac{-0.482}{3.644} \right)^2 \cdot (0.018)} \right]^{-1/2} \cdot 100\%$$

$$= 3.74\%\,\mathrm{FS} \text{ five-bits interpolated video}$$

PROBLEMS

6-1. A 1 V rms full-scale signal of 1-Hz BW in 10 mV rms of random interference is applied to an A/D converter with an f_{hi} of 1 KHz being sampled at an f_s of 10 Hz. Determine the random aliasing error in %FS in the absence of a presampling filter with the aid of Equations (6-8) to (6-10).

6-2. Consider a full-scale amplitude signal of 1-Hz BW at an f_s of 10 Hz recovered by direct-D/A conversion. Compare the intersample error of this signal both by Equation (6-13) and application of the D/A output intepolator function in Table 6-3. Repeat the intersample error evaluation for this signal considering a D/A + three-pole Butterworth output interpolator function.

6-3. A compact disc player employs a 44.1 KHz sample rate with 14-bit quantization and a linear interpolator for output signal reconstruction. For typical signal amplitudes of 30% full scale, evaluate interpolated output intersample error and binary accuracy equivalents at signal frequencies of 120 Hz, 1.2 KHz, and 12 KHz.

REFERENCES

1. Davenport, W.B., Jr. and Root, W.L., *An Introduction to the Theory of Random Signals and Noise,* New York: McGraw-Hill, 1958.
2. Gardenshire, L.W., "Selecting Sample Rates," *ISA Journal,* April 1964.
3. Garrett, P.H., *High Perfomance Instrumentation and Automation,* CRC Press, 2005.
4. Hoffman, A.W., *Interpolative Digital-to-Analog Converter for Computer Output Systems,* BSEET Sr. Thesis, Univ. of Cincinnati, 1984.
5. Jerri, A.J., "The Shannon Sampling Theorem—Its Various Extensions and Applications: A Tutorial Review," *Proceedings of the IEEE,* Vol. 65, No. 11, November 1977.
6. Kolmogoroff, A., "Interpolation and Extrapolation von Stationaren Zufalligen Folgen," *Bulletin Academic Sciences, Serial Mathematics,* Vol. 5, 1941 (USSR).
7. Nyquist, H., "Certain Topics in Telegraph Transmission Theory," *Transactions of the AIEE,* Vol. 47, February 1928.
8. Papoulis, A., *Probability, Random Variables, and Stochastic Processes,* New York: McGraw-Hill, 1965.
9. Raemer, H.R., *Statistical Communication Theory and Applications,* Englewood Cliffs, NJ: Prentice-Hall, 1969.
10. Schwartz, M., Bennett, W.R., and Stein, S., *Communications Systems and Techniques,* New York: McGraw-Hill, 1966.
11. Shannon, C.E. and Weaver, W., *The Mathematical Theory of Communication,* Urbana, IL: University of Illinois Press, 1949.
12. Whittaker, E.T., "On Functions Which Are Represented by the Expansions of the Interpolation Theory," *Proceedings of the Royal Society,* Vol. 35, 1915.
13. Wiener, N., *Extrapolation, Interpolation, and Smoothing of Stationary Time Series with Engineering Applications,* Cambridge, MA: MIT Press, 1949.

CHAPTER 7

Instrumentation Analysis Suite, Error Propagation, Sensor Fusion, and Interfaces

7-0 INTRODUCTION

The preceding chapters have demonstrated comprehensive end-to-end modeling of sensor data acquisition through signal conditioning and data conversion functions, including output signal reconstruction and actuation. These models beneficially provide a physical description of instrumentation performance with regard to device and system choices to verify fulfillment of the measurement accuracy required. Modeled instrumentation system error, therefore, valuably permits performance to be quantitatively forecast prior to implementation for measurement confidence, including instrumented process system designs.

This chapter integrates complete multisensor instrumentation systems, error propagation, and sensor fusion from industrial and aerospace applications. Notable is the development of an analysis suite spreadsheet for comprehensive design analysis, such as utilized in avionics systems to achieve more critical performance accountability than is normally required for commercial applications. A final section introduces instrumentation categories, including their bus and network interfaces.

7-1 AEROSPACE COMPUTER I/O DESIGN WITH ANALYSIS SUITE

Computer-based instrumentation is widely employed to interface analog measurement signals to digital systems. It is common for applications to involve joint input/output operations, by which analog signals are also recovered for actuation or end-use purposes following digital processing. Instrumentation error models derived for devices and transfer functions in the preceding chapters will be assembled into an instrumentation analysis suite for I/O system design. This workbook demonstrates evaluating the cumulative error of conditioned and converted sensor signals input to a computer data bus including their output reconstruction in analog form, with the option for substituting alternate circuit topologies and devices for system optimization. This is especially of value for appraising I/O products in implementation selection.

Figure 7-1 describes an aircraft engine test cell vibration analyzer I/O system combining the signal conditioning example of Figure 4-6 with Datel data converter devices interfacing a tunable digital band-pass filter for frequency resolution of vibration amplitude signals. Signal conditioning includes a premium performance acquisition channel consisting of a 0.1%FS systematic error piezoresistive bridge-strain-gauge accelerometer that is biased by isolated ± 0.5 V dc regulated excitation and connected differentially to an Analog Devices AD624C preamplifier accompanied by up to 1 V_{rms} of common-mode random noise. The harmonic sensor signal has a maximum amplitude of 70 mV_{rms}, corresponding to $\pm 10g$, up to 100 Hz fundamental frequencies with a first-order rolloff to 7 mV_{rms} at a 1 KHz bandwidth. The preamplifier differential gain of 50 raises this signal to a ± 5 V_{pp} full-scale value while attenuating random interference, in concert with the presampling filter, to 0.006%FS signal quality or 212 μV output rms (from ± 5 $V_{pp}/\sqrt{2}$ times 0.00006 numerical). The associated sensor-loop internal noise of 15 μV_{pp} plus preamplifier referred-to-input errors total 27μV dc with reference to Table 4-5. This defines a signal dynamic range of $\sqrt{2} \cdot 70$ $mV_{rms}/27$ μV, or 71 voltage dB, approximating 12 bits of amplitude resolution. Amplitude resolution is not further limited by subsequent system devices, which actually exceed this performance, such as the 16-bit data converters.

It is notable that the Butterworth low-pass presampling signal-conditioning filter achieves signal quality upgrading for random noise through a linear-filter approximation to matched-filter efficiency by the provisions of Chapter 4, Section 4-2. This filter also provides undersampled random noise aliasing attenuation above f_0 from Chapter 6, Section 6-2 with cutoff frequency derating to minimize its mean filter error from Chapter 3, Section 3-3. Errors associated with the amplifiers, S/H, AMUX, A/D, and D/A data converters are primarily nonlinearities and temperature-drift contributions that result in LSB equivalents between 9–13 bits of accuracy.

The A/D and D/A converters are also elected to be discrete-switching devices to avoid signal artifacts possible with sigma–delta type converters. Sample rate f_s,

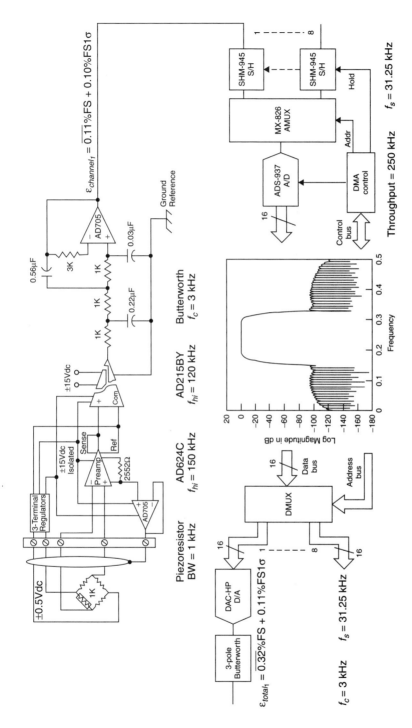

Figure 7-1. Aircraft engine vibration analyzer I/O design.

determined by dividing the available 250 KHz DMA-transfer rate between eight channels, is 31 times the 1 KHz signal BW. That provides excellent sampled-data performance in terms of small sinc error, negligible noise aliasing of the 212 μV_{rms} of residual random interference, and accurate output signal recovery.

Figure 7-2 shows the error of converted input signal versus frequency applied to a digital data bus, where its step-interpolated intersample error value is the dominant contributor of 0.63%FS at full bandwidth. The combined total input error of 0.83%FS remains constant from 10% of signal bandwidth to the 1 KHz full-bandwidth value, owing to harmonic signal amplitude rolloff with increasing frequency, declining to 0.32%FS at 1% bandwidth. It is significant that the sample-image-frequency spectra described in Chapter 6, Section 6-4 are regenerated by each I/O sampling operation from S/H through D/A converter devices, and that these spectra are transformed with signal transfer from device to device when there is a change in f_s. Increasing f_s accordingly results both in sampled-image-frequency spectra being heterodyned to higher frequencies and a decreased mean signal at-

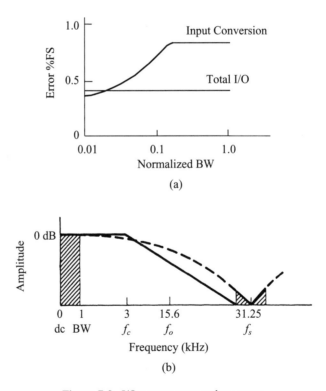

Figure 7-2. I/O system error and spectrum.

tenuation from the associated sinc function. This describes the basis of oversampling, defined as sampling rates greater than the Nyquist f_s/BW ratio of two, which offers enhanced output reconstruction through improved attenuation of the higher sampled-image-frequency spectra by the final postfiltering interpolator.

The illustrated I/O system and its accompanying analysis suite example provide detailed accountability of total system performance and realize the end-to-end goal of not exceeding 0.1%FS error for any contributing element to the error summary of Table 7-1. Output signal reconstruction is effectively performed by a post-D/A Butterworth third-order low-pass filter derated to reduce its component error while simultaneously minimizing intersample error. This implementation results in an ideal flat total instrumentation channel error versus bandwidth, shown in Figure 7-1, of 0.43%FS. This error is equivalent to approximately eight bits of true amplitude accuracy within 12 bits of signal dynamic range and 16 bits of data quantization.

The Microsoft Excel Integrated Instrumentation Analysis Suite Spreadsheet contains an interactive workbook of complete instrumentation system error models in 69 Kbytes for computer-assisted engineering design. The first page of this six-sheet analysis suite permits defining sensor and excitation input values, including signal bandwidth and differential signal voltage amplitude, and provides for both random and coherent interference accommodation. This data is utilized

Table 7-1. I/O Instrumentation summary

Element	$\varepsilon_{\%FS}$	Comment
Sensor	0.100000	Piezoresistor bridge
Interface	0.010000	Residual differential excitation
Amplifiers	0.033950	AD624C + AD215BY
Presampling filter	0.115000	Three-pole Butterworth
Signal quality	0.006023	Random noise ε_{rand}
Noise aliasing	0.000004	212 μV residual interference
Sinc	0.084178	Average signal attenuation
Multiplexer	0.004001	Average transfer error
Sample hold	0.020633	400 ns acquisition time
A/D	0.002442	16-bit subranging
D/A	0.013032	16-bit converter
Interpolator filter	0.1150000	Three-pole Butterworth
Intersample	0.000407	Output interpolation
ε_{total}	0.318179%FS	$\Sigma\overline{mean}$
	0.109044%FS	1σ RSS
	0.427223%FS	$\Sigma\overline{mean} + 1\sigma$ RSS

for subsequent model calculations, and returns the input signal-to-noise ratio and required system voltage gain. Values for the vibration I/O system shown in Figure 7-1 are exampled throughout the pages of this spreadsheet. Specific sensor and excitation input values and model calculations associated with this example are presented in greater detail surrounding Figure 4-6.

The second and third pages accommodate up to four cascaded amplifiers per system, whereby 13 parameters are entered for each amplifier selected from manufacturer's data. Seven additional quantities related to sensor-circuit parameters are also required, which ordinarily accompany only the front-end amplifier in a system. Seven calculated equivalent-input-error voltages are returned for each amplifier, defining their respective error budgets, and combined in an eighth amplifier value expressing error as %FS. A detailed representation of the model calculations for amplifiers employed in this example are tabulated in Tables 4-3 and 4-4.

The fourth page evaluates linear signal conditioning performance in terms of attained signal quality, including specification of parameters for a band-limiting presampling filter, which serves a dual role in signal conditioning and aliasing prevention. Calculated values returned include residual coherent and random interference error as well as filter device error from Chapter 3, Section 3-3. Sampled data parameters including a trial sample rate f_s, and undersampled coherent and random interference amplitude values existing above the Nyquist frequency ($f_s/2$), are then entered so that aliasing error may be evaluated. The amplitude values are proportional to postsignal-conditioning residual errors, evaluated by Equations (4-15) and (4-16), as determined by scaling to the system full-scale voltage value. Returned values of aliasing error and sinc error rely upon corresponding models developed in Chapter 6.

Remaining analysis suite spreadsheet entries consist of parameter values obtained from manufacturers for modeling five data converter devices, including AMUX, S/H, A/D, D/A, and output interpolator devices. The combined error for all of the device and system contributions are automatically tabulated for full signal BW in Figure 7-2 including output interpolation, and plotted from 1% to 100% of signal BW both for converted signals on a computer data bus and input to output with choice of output interpolator device. Available modeled data reconstructors include direct-D/A, one-pole RC, and three-pole Butterworth interpolators.

Sensor
1 kΩ piezoresistor bridge with $F = $ mA response at 0.1%FS error

Interface

Sensor voltage excitation ± variation
± 0.5 V dc ± 50 μV or 0.01%FS

Signal quality

$$\varepsilon_{random} = \frac{V_{cm}}{V_{diff}} \cdot \left[\frac{R_{diff}}{R_{cm}} \right]^{1/2} \cdot \frac{A_{V_{cm}}}{A_{V_{diff}}} \cdot \left[\frac{2}{k} \frac{f_c}{f_{hi}} \right]^{1/2} \cdot 100\% \qquad (4\text{-}16)$$

$$= \frac{1\,V}{7\,mV} \cdot \left[\frac{1\,G\Omega}{1\,G\Omega} \right]^{1/2} \cdot \frac{10^{-4}}{50} \cdot \left[\frac{2}{0.9} \frac{3\,kHz}{150\,kHz} \right]^{1/2} \cdot 100\%$$

$$= 0.006023\%FS$$

Sample hold	
Acquisition error	0.00760%
Nonlinearity	0.00040
Gain	0.02000
Temperature coefficient	0.00500
$\varepsilon_{S/H}$	0.020633%FS 1 σ RSS

Amplifiers		
Parameter	AD624C	AD215BY
V_{os}	Trimmed	Trimmed
$\dfrac{dV_{os}}{dT} \cdot dT$	2.5 μV	20 μV
$I_{os} \cdot R_i$	$\overline{10\ \mu V}$	$\overline{15\ \mu V}$
V_{Npp}	15 μV	2 μV
$f(Av) \cdot \dfrac{V_{O_{FS}}}{A_{v_{diff}}}$	$\overline{1\ \mu V}$	$\overline{250\ \mu V}$
$\dfrac{dAv}{dT} \cdot dT \cdot \dfrac{V_{O_{FS}}}{A_{v_{diff}}}$	5 μV	750 μV
$\varepsilon_{ampl\ RTI}$	$(\overline{11} + 16)\mu V$	$(\overline{265} + 750)\mu V$
$\varepsilon_{ampl\ \%FS}$	0.027%FS	0.020%FS

Analog multiplexer	
Transfer error	$\overline{0.003}\%$
Leakage	0.001
Crosstalk	0.00005
$\varepsilon_{\text{AMUX}}$	$\overline{0.004001}\%\text{FS}$ $\quad$ $\Sigma\overline{\text{mean}} + 1\,\sigma\,\text{RSS}$

Noise aliasing

$$\varepsilon_{\text{random alias}} = \frac{\sqrt{2}\ 100\%}{\left[\text{SNR}_{\text{random alias}}\right]^{1/2}} \tag{6-10}$$

$$= \frac{\sqrt{2}\ 100\%}{\left[V_{\text{FS}}^2 / (V_{\text{noise rms}})^2 \cdot \left(\dfrac{1}{\left[1 + \left(\dfrac{f_s}{f_c} \right)^{2n} \right]^{1/2}} \right)^2 \right]^{1/2}}$$

$$= \frac{\sqrt{2}\ 100\%}{\left[V_{\text{FS}}^2 / \left(\dfrac{\sqrt{2} \cdot 1\ V_{\text{rms}} \cdot 10^{-4}}{5\ \text{V}}\ V_{\text{FS}} \right)^2 \cdot \left(\dfrac{1}{\left[1 + \left(\dfrac{31.25\ \text{kHz}}{3\ \text{kHz}} \right)^6 \right]^{1/2}} \right)^2 \right]^{1/2}}$$

$$= \frac{\sqrt{2}\ 100\%}{\left[V_{\text{FS}}^2 / (9 \times 10^{-10}\ V_{\text{FS}}^2)(0.78 \times 10^{-6}) \right]^{1/2}}$$

$$= \frac{\sqrt{2}\ 100\%}{\left[V_{\text{FS}}^2 / 7 \times 10^{-16}\ V_{\text{FS}}^2 \right]^{1/2}}$$

$$= 0.000004\%\text{FS}$$

Sinc

$$\varepsilon_{\text{NRZ sinc}} = \frac{1}{2}\left(1 - \frac{\sin \pi BW / f_s}{\pi BW / f_s} \right) \cdot 100\% \tag{6-13}$$

$$= \frac{1}{2}\left(1 - \frac{\sin \pi\ 1\ \text{kHz} / 31.25\ \text{kHz}}{\pi\ 1\ \text{kHz} / 31.25\ \text{kHz}} \right) \cdot 100\%$$

$$= \overline{0.084178}\%\text{FS}$$

16-Bit A/D		
Mean integral nonlinearity (1LSB)	$\overline{0.0011}\%$	
Noise + distortion	0.0001	
Quantizing uncertainty ($\frac{1}{2}$LSB)	0.0008	
Temperature coefficient	0.0011	
$\varepsilon_{\text{A/D}}$	0.002442%FS	$\Sigma\overline{\text{mean}} + 1\,\sigma\,\text{RSS}$

16-Bit D/A		
Mean integral nonlinearity (1LSB)	$\overline{0.003}\%$	
Noise + distortion	0.0008	
Temperature coefficient	0.0100	
$\varepsilon_{\text{D/A}}$	0.013032%FS	$\Sigma\overline{\text{mean}} + 1\,\sigma\,\text{RSS}$

Interpolated intersample

$$\varepsilon_{\Delta V} = \left[\frac{V_{o\,\text{FS}}^2}{V_s^2 \left\{ \text{sinc}^2\left(1 - \frac{BW}{f_s}\right)\left[1 + \left(\frac{f_s - BW}{f_c}\right)^{2n}\right]^{-1} + \text{sinc}^2\left(1 + \frac{BW}{f_s}\right)\left[1 + \left(\frac{f_s + BW}{f_c}\right)^{2n}\right]^{-1}\right\}} \right]^{-1/2} \cdot 100\% \qquad \text{(Table 6-3)}$$

$$= \left[\frac{(10\,V)^2}{(1V)^2 \cdot \left\{ \text{sinc}^2\left(1 - \frac{1\,\text{kHz}}{31.25\,\text{kHz}}\right)\left[1 + \left(\frac{31.25\,\text{kHz} - 1\,\text{kHz}}{1\,\text{kHz}}\right)^{6}\right]^{-1} + \text{sinc}^2\left(1 + \frac{1\,\text{kHz}}{31.25\,\text{kHz}}\right)\left[1 + \left(\frac{31.25\,\text{kHz} + 1\,\text{kHz}}{1\,\text{kHz}}\right)^{6}\right]^{-1}\right\}} \right]^{-1/2} \cdot 100\%$$

$$= 0.000407\%\text{FS}$$

INTEGRATED INSTRUMENTATION ANALYSIS SUITE SPREADSHEET

Parameter	Symbol	Value	Units	Comment (unshaded: entered; shaded: calculated)
		Sensor and excitation entries		
Sensor error type	ε	S	M or S	M = Static Mean, S = Variable Systemic
Sensor error value	$\varepsilon_{\text{sensor}}$	0.1	%FS	Sensor full-scale error from manufacturer's information
Peak input signal voltage	V_s	0.1	Volts	Sensor full-scale signal voltage DC or RMS $\sqrt{2}$ up to fundamental or BW/10 for harmonic signals
Peak output signal voltage	$V_{o\text{FS}}$	5.00	Volts	System full-scale voltage DC or RMS $\sqrt{2}$
Signal bandwidth	BW	1000	Hertz	Sensor signal bandwidth to highest frequency of interest
Interface error type	ε	S	M or S	M = Static Mean, S = Variable Systemic
Common mode interference	V_{cm}	1.0	Volts	Input common-mode DC or RMS random and/or coherent interference amplitude
Differential input signal @ BW	V_{diff}	0.007	Volts	Sensor DC or RMS differential voltage at full BW for signal quality evaluation
Coherent interference present	Coh	N	Y or N	Enter N if no coherent interference
Coherent interference frequency	f_{coh}	0	Hertz	Frequency of coherent interfering signal if present
Random interference present	Rand	Y	Y or N	Enter a Y if random noise is present, N if not
Input interface error	$\varepsilon_{\text{interface}}$	0.01	%FS	Interface termination or sensor excitation error
Sinusoidal/harmonic	H or S	H		Enter H for complex harmonic signals and S for sinusoidal or DC signals
Required voltage	A_v	50	V/V	$V_{o\text{FS}}/V_S$ total gain between sensor and A/D converter
Input SNR	SNR_i	4.900E-05	(V/V)^2	Input signal-to-noise ratio as $(V_{\text{diff}}/V_{\text{cm}})^2$

| | | \multicolumn{4}{c}{Amplifier data} | | |
| | | \multicolumn{4}{c}{Amplifier error budget parameters} | | |
Parameter	Symbol	Amp_1	Amp_2	Amp_3	Amp_4	Units	Comment
Amplifier type		AD624C	AD215BY				Manufacturer's part number
Common mode resistance	R_{cm}	1.00E+09	5.00E+09			Ohms	Input common mode resistance
Differential resistance	R_{diff}	1.00E+09	1.00E+12			Ohms	Input differential resistance
Amplifier cutoff frequency	f_{hi}	150,000	120,000			Hertz	Amplifier −3 dB cutoff frequency
Mean offset voltage amplitude	V_{OS}	2.500E-05	4.000E-04			Volts	DC voltage between differential inputs
Voltage offset temperature drift	$\Delta V_{OS}/\Delta T$	2.50E-07	2.00E-06			Volts/°C	Input offset voltage temperature drift
Temperature variation	ΔT	10	10			°C	Circuit temperature variation
Offset current	I_{OS}	0.01	0.3			μA	DC input offset bias current difference
Current offset temperature drift	$\Delta I_{OS}/\Delta t$	2.00E-05	1.00E-03			μA/C°	Input offset current temperature drift
Ambient temperature	T_{amb}	20	20			°C	Temperature of system environment
Sensor current amplitude	I_{DC}	1000	0			μA	Sensor DC current flow if present
Contact noise frequency	$f_{contact}$	100	100			Hertz	Contact noise test frequency (convention 10% of BW)
Offset voltage nulled	V_{OSNull}	N	N			A or N	Enter A if V_{OS} added to ε_{amp}, N if nulled
Input noise voltage equivalent	V_n	0.004	0			μV/$\sqrt{Hz}$	Amplifier RMS noise voltage per root Hertz
Input noise current equivalent	I_n	0.06	0			pA/$\sqrt{Hz}$	Amplifier RMS noise current per root Hertz
Common mode rejection	CMRR	5.00E+05	1.00E+05			V/V	Numeric common-mode rejection ratio

Parameter	Symbol	Amplifier error budget parameters				Units	Comment
		Amp_1	Amp_2	Amp_3	Amp_4		
Gain nonlinearity	$f(A_V)$	10	50			ppm	Gain nonlinearity over gain range
Peak output signal voltage	V_{oFS}	5	5			Volts	Amplitude full-scale output voltage DC or RMS $\cdot \sqrt{2}$
Differential gain	$A_{V_{\text{diff}}}$	50	1			V/V	Closed-loop differential gain
Gain temperature drift	$\Delta A_v/\Delta T$	5	15			ppm/C°	Gain temperature drift
Source resistance	R_s	1000	50			Ohms	Source resistance seen by respective amplifier
Voltage drift from temp.	ΔV_{OS}	2.500E-06	2.000E-05	0.000E+00	0.000E+00	Volts	Input offset voltage temperature drift
Mean offset I_{OS} voltage	$I_{OS}R_s$	1.000E-05	1.500E-05	0.000E+00	0.000E+00	Volts	Voltage error due to input offset current
Thermal noise	V_t	4.022E-09	8.993E-10	0.000E+00	0.000E+00	V/$\sqrt{\text{Hz}}$	Thermal RMS noise in sensor circuit
Contact noise	V_c	1.802E-09	0.000E+00	0.000E+00	0.000E+00	V/$\sqrt{\text{Hz}}$	Contact RMS noise in sensor circuit
Total noise	V_{Npp}	1.521E-05	2.056E-06	0.000E+00	0.000E+00	Volts	$6.6\text{RSS}(V_t + V_c + V_n)\sqrt{f_{\text{hi}}}$
Mean gain nonlinearity	$V_{f(A_V)}$	1.000E-06	2.500E-04	0.000E+00	0.000E+00	Volts	Voltage error due to gain nonlinearity
Gain temperature drift	$V_{\Delta A_V/\Delta T}$	5.000E-06	7.500E-04	0.000E+00	0.000E+00	Volts	Voltage error due to gain temperature drift
Amplifier errors	ε_{amp}	0.02721	0.02031	0.00000	0.00000	%FS	(Σ mean V + RSS other V) $\cdot$ ($A_{V_{\text{diff}}}/V_{oFS}$) $\cdot$ 100%

Signal quality and presampling filter entries				
Parameter	Symbol	Value	Units	Comment (unshaded: entered; shaded: calculated)
Presampling filter	n	3	Poles	Valid for 1–8 Butterworth poles for harmonic signals
Filter present		Y	Y or N	Enter N if no filter present

Parameter	Symbol	Value	Units	Comment (unshaded: entered; shaded: calculated)
Filter efficiency	K	0.9	Parameter	Linear filter efficiency relative to matched filtering (0.9 default value)
Mean filter error	$\varepsilon_{\text{filter}}$	0.115	%FS	Presampling filter error for complex harmonic signal
Filter cutoff frequency	f_c	3000	Hertz	Presampling filter cutoff frequency
Amplifier SNR	SNR_{amp}	1.23E+07	W/W	Signal conditioning amplifier output signal-to-noise power ratio
Amplifier SNR error	$\varepsilon_{\text{amp SNR}}$	0.02857	%FS	Signal conditioning amplifier output error
Coherent filter SNR	SNR_{coh}	—	W/W	Filter output coherent signal-to-noise ratio as power ratio
Coherent filter SNR error	$\varepsilon_{\text{coh amp}}$	—	%FS	Filter output full-scale signal error for filter SNR
Random filter SNR	SNR_{rand}	5.5E+08	W/W	Random filter SNR
Random filter SNR error	$\varepsilon_{\text{rand amp}}$	0.00602	%FS	Filter output full-scale signal error for filter SNR (random)
Total signal quality	ε_{sq}	0.00602	%FS	$\varepsilon_{\text{amp SNR}}$ or RSS ($\varepsilon_{\text{rand amp}} + \varepsilon_{\text{coh amp}}$) with filter

Aperture, sinc, and aliasing entries

Parameter	Symbol	Value	Units	Comment (unshaded: entered; shaded: calculated)
Aperture time	t_a	0.002	μs	Aperture time of sample and hold
Sample rate	f_s	31250	Hertz	Sample rate selected
Undersampled coherent interference	A_{coh}	0	V RMS	Amplitude of the coherent undersampled RMS noise at S/H and A/D
Undersampled random interference	A_{rand}	2.12E-04	V RMS	Amplitude of the random undersampled RMS noise at S/H and A/D
Coherent alias frequency	$f_{\text{coh alias}}$	0	Hertz	Undersampled coherent aliasing source frequency at input
Interfering baseband alias	f_{alias}	0	Hertz	Baseband coherent aliasing frequency
Mean aperture error	ε_a	3.290E-10	%FS	Aperture error as percent full scale
ZOH intersample error	$\varepsilon_{\Delta VZOH}$	0.6358	%FS	ZOH intersample error at full BW prior to interpolation
Mean sinc error	$\varepsilon_{\text{NRZ sinc}}$	0.0842	%FS	Average sinc error for NRZ sampling

Parameter	Symbol	Value	Units	Comment (unshaded: entered; shaded: calculated)
Coherent alias error	$\varepsilon_{\text{coh alias}}$	0.00E+00	%FS	Aliasing by the undersampled coherent interference amplitude
Random alias error	$\varepsilon_{\text{rand alias}}$	7.50E-06	%FS	Aliasing by the undersampled random interference amplitude
Total alias error	$\varepsilon_{\text{alias}}$	7.50E-06	%FS	$\text{RSS}(\varepsilon_{\text{coh alias}} + \varepsilon_{\text{rand alias}})$

Multiplexer entries

Parameter	Symbol	Value	Units	Comment (unshaded: entered; shaded: calculated)
Mean transfer error	$\varepsilon_{\text{trans}}$	0.003	%FS	Mean transfer error as percent full scale
Crosstalk	$\varepsilon_{\text{cross}}$	0.00005	%FS	Crosstalk error as percent full scale
Leakage	$\varepsilon_{\text{leak}}$	0.001	%FS	Leakage error as percent full scale
Mean multiplexer error	$\varepsilon_{\text{AMUX}}$	0.00400	%FS	

Sample-hold entries

Parameter	Symbol	Value	Units	Comment (unshaded: entered; shaded: calculated)
Acquisition error	ε_{acq}	0.00076	%FS	Acquisition error following required settling time
Nonlinearity	ε_{lin}	0.0004	%FS	Sample-hold nonlinearity errors
Gain	$\varepsilon_{\text{gain}}$	0.02	%FS	Gain errors
Temperature coefficient	$\varepsilon_{\text{tempco}}$	0.005	%FS	Temperature coefficient errors
Sample-hold error	$\varepsilon_{\text{S/H}}$	0.02063	%FS	RSS sample hold entries

Analog-to-digital converter entries

Parameter	Symbol	Value	Units	Comment (unshaded: entered; shaded: calculated)
A/D word length	Data Bus	16	Bits	Converter wordlength
Quantizing uncertainty	ε_{q}	0.0008	%FS	Quantizing uncertainty as %FS of ½ LSB
Mean integral nonlinearity	ε_{INL}	0.0011	%FS	Mean integral nonlinearity as %FS
Noise + distortion	$\varepsilon_{\text{N+D}}$	0.0001	%FS	Noise plus distortion as a %FS
Temperature coefficient	$\varepsilon_{\text{tempco}}$	0.0011	%FS	Temperature coefficient errors
A/D error	$\varepsilon_{\text{A/D}}$	0.00244	%FS	$\varepsilon_{\text{INL}} + \text{RSS other}$

Digital-to-analog converter entries

Parameter	Symbol	Value	Units	Comment (unshaded: entered; shaded: calculated)
D/A word length	Data Bus	16	Bits	Converter word length
Mean integral nonlinearity	ε_{INL}	0.003	%FS	Mean integral nonlinearity as %FS
Temperature coefficient	ε_{tempco}	0.01	%FS	Temperature coefficient errors
Noise + distortion	ε_{N+D}	0.0008	%FS	Noise plus distortion as a %FS
D/A error	$\varepsilon_{D/A}$	0.01303	%FS	ε_{INL} + RSS other

Data reconstruction entries

Parameter	Symbol	Value	Units	Comment (unshaded: entered; shaded: calculated)
Reconstructor type	Recovery	B	D, R, or B	Type of reconstruction circuit (D = direct D/A, R = 1-pole RC, B = Butterworth 3-pole filter)
Interpolator poles	N	3	Poles	0 for direct D/A, 1 for 1-pole RC, 3 for Butterworth
Interpolated intersample error	$\varepsilon_{\Delta V}$	0.0004	%FS	Interpolated intersample error
Mean interpolator device error	ε_{interp}	0.1150	%FS	Interpolator device error
Reconstruction system error	ε_{recov}	0.1280	%FS	ε_{interp} + RSS($\varepsilon_{D/A}$ + $\varepsilon_{\Delta V}$)
Cutoff frequency	f_{cr}	3000	Hertz	Interpolator device filter cutoff frequency

7-2   MEASUREMENT ERROR PROPAGATION IN EXAMPLE AIRFLOW PROCESS

Figure 7-3 describes a measurement process employed in turbine engine manufacture for determining blade internal airflows, with respect to design requirements, essential to part heat transfer and rogue blade screening. A preferred evaluation method is to describe blade airflow in terms of fundamental geometry such as its effective flow area. The implementation of this measurement process is described by analytical Equations (7-1) and (7-2), where uncontrolled air density ρ appears as a ratio to effect an air-density-independent airflow measurement. That outcome beneficially enables quantitative determination of part airflows from known parameters and pressure measurements that are defined in Table 7-2. The airflow process mechanization accordingly consists of two plenums with specific volumetric airflows and four pressure measurements. Appendix A describes the derivation of analytical equations upon which this process is based.

In operation, the fixed and measured quantities determine part flow area employing two measurement sequences. Plenum volumetric airflows are initially reconciled for pitot-stagnation-pressures $P_0 - P_0$, thereby obtaining the plenums ratio of internal airflow velocities V_{p1}/V_{r1}. The quantities are then arranged into a ratio of plenum volumetric airflows, which, combined with gauge and differential pressure measurements P_{r1}, P_{atm}, and $P_{p1} - P_{r1}$, permit expression of air-density-independent part flow area A_{P2} by Equation (7-2). Equation (7-3) then describes multisensor error propagation determined from analytical process Equations (7-1) and (7-2) with the aid of Table 7-3. Part flow area error results from the algorithmic propagation of four independent pressure sensor instrumentation errors in this two-sequence measurement example, where individual sequence errors are summed because of the absence of correlation between the two measurements [1].

In the first sequence, an equalized pitot pressure measurement ΔP_0 is acquired, defining Bernoulli's Equation (7-1). The algorithmic influence of this pressure measurement is represented by the sum of its static mean plus single RSS error contribution in the first sequence of Equation (7-3). The second measurement sequence is defined by Equation (7-2), whose algorithmic error propagation is obtained from arithmetic operations on measurements P_{r1}, P_{atm}, and

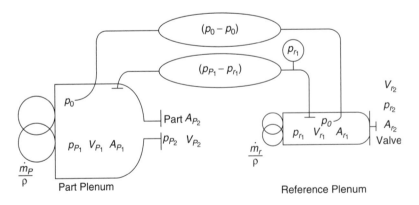

Figure 7-3. Part airflow measurement process.

Table 7-2. Airflow process parameter glossary

Known airflow process parameters			Measured airflow process parameters		
Symbol	Value	Description	Symbol	Value	Description
$\dfrac{\dot{m}_r}{\rho}$	$\dfrac{\text{ft}^3}{\text{min}}$	Reference plenum volumetric flow	A_{p2}	ft^2	Part effective flow area
A_{r_1}	ft^2	Reference plenum inlet area	$P_{p_1} - P_{r_1}$	lb/ft^2	Part-to-reference plenum differential pressure
V_{r_1}	$\dfrac{\text{ft}^3}{\text{min}}$	Reference plenum inlet velocity	P_{r_1}	lb/ft^2	Reference plenum-gage pressure
A_{p_1}	ft^2	Part plenum inlet area	$P_0 - P_0$	lb/ft^2	Reference-part plenum equalized stagnation pressures
ρ	$0.697\text{E-}6$ $\dfrac{\text{lb}-\text{min}}{\text{ft}^4}$	Air density at standard temperature and pressure	P_{p2}	$P_{\text{atm}}\ \text{lb/ft}^2$	Part plenum exit pressure

P_{p1}-P_{r1} represented by the sum of their mean plus RSS error contributions in Equation (7-3). For the first sequence of Equation (7-3), only the differential-pitot-stagnation-pressure measurement $P_0 - P_0$ is propagated as algorithmic error. In the following second sequence, part plenum inlet area A_{p1}, air density ρ, and reference plenum inlet velocity V_{r1} all are constants that do not appear as propagated error. However, the square-root exponent influences the error of the three

Table 7-3. Instrumentation error algorithmic propagation

Instrumentation error	Algorithmic operation	Error influence
$\varepsilon\ \overline{\text{mean}}$ %FS	Addition	$\Sigma\ \overline{\varepsilon}_{\text{mean}}$ %FS
	Subtraction	$\Sigma\ \overline{\varepsilon}_{\text{mean}}$ %FS
	Multiplication	$\Sigma\ \overline{\varepsilon}_{\text{mean}}$ %FS
	Division	$\Sigma\ \overline{\varepsilon}_{\text{mean}}$ %FS
	Power function	$\Sigma\ \overline{\varepsilon}_{\text{mean}}$ %FS $\times$ |exponent value|
$\varepsilon_{\%\text{FS}\ 1\sigma}$	Addition	RSS ε %FS 1σ
	Subtraction	RSS ε %FS 1σ
	Multiplication	RSS ε %FS 1σ
	Division	RSS ε %FS 1σ
	Power function	RSS ε %FS $1\sigma \times$ |exponent value|

pressure measurements included in Equation (7-2) by the absolute value shown. Four independent nine-bit accuracy pressure measurements are accordingly combined by these equations to realize an eight-bit accuracy part flow area. Equation (7-4) defines the error for each pressure measurement, ε_{sensor}, including its signal conditioning and data conversion at the host computer data bus by the methods of the preceding chapters. Equation (7-4) describes the measurement error for each of the four pressure measurements.

$$\Delta P_0 = (P_{P_1} + \tfrac{1}{2}\rho V_{P_1}^2) - (P_{r_1} + \tfrac{1}{2}\rho V_{r_1}^2) \qquad P_0 \text{ equilibrium sequence} \qquad (7\text{-}1)$$

$$A_{P2} = A_{P1} \cdot \left[\frac{\rho - 2(P_{P_1} - P_{r_1})/V_{r_1}^2}{\rho + 2(P_{r_1} - P_{atm})/V_{r_1}^2} \right]^{1/2} \qquad \text{part flow area sequence} \quad (7\text{-}2)$$

$$\varepsilon_{\Delta P0} + \varepsilon_{AP2} = \left\{ \overline{\varepsilon_{\text{mean } \Delta P_0}} \,\%FS + \varepsilon_{\Delta P_0} \,\%FS1\sigma \right\} \qquad \text{1st sequence} \qquad (7\text{-}3)$$

$$+ \left\{ \left| \tfrac{1}{2} \right| \left[\overline{\varepsilon_{\text{mean } \Delta P_{1^{-r_1}}}} + \overline{\varepsilon_{\text{mean } P_{r_1}}} + \overline{\varepsilon_{\text{mean } P_{atm}}} \right] \%FS \right.$$

$$+ \left. \left| \tfrac{1}{2} \right| \left[\varepsilon_{\Delta P_{1^{-r_1}}}^2 + \varepsilon_{P_{r_1}}^2 + \varepsilon_{P_{atm}}^2 \right]^{1/2} \%FS1\sigma \right\} \qquad \text{2nd sequence}$$

$$= \left\{ \overline{0.1\%}FS + 0.1\%FS1\sigma \right\} \qquad \text{1st sequence}$$

$$+ \left\{ \left| \tfrac{1}{2} \right| \left[\overline{0.1} + \overline{0.1} + \overline{0.1} \right] \%FS \right.$$

$$+ \left. \left| \tfrac{1}{2} \right| \left[0.1^2 + 0.1^2 + 0.1^2 \right]^{1/2} \%FS \, 1\sigma \right\} \qquad \text{2nd sequence}$$

$$= \overline{0.25}\%FS + 0.186\%FS \, 1\sigma \qquad \text{8-bit accuracy}$$

$$\varepsilon_{sensor} = \overline{0.1}\%FS + 0.1\,\%FS1\sigma \qquad \text{9-bit accuracy} \qquad (7\text{-}4)$$

7-3   HOMOGENEOUS AND HETEROGENEOUS SENSOR FUSION

Preceding sections have demonstrated comprehensive error modeling of instrumentation system designs for both simultaneous and sequential data acquisition applications. This section addresses homogeneous and heterogeneous multisensor fusion architectures in decision and control applications. Sonar signal processing, shown by Figure 7-4, illustrates the basics of homogeneous sensor fusion. That includes signal conditioning of redundant sensors for noise reduction, concluding in a data fusion display providing measured property retrieval and ambiguity reduction unavailable with single sensors. Since random noise constitutes the principal additive uncertainty to sonar signals, whose sensors possess only nominal mean error, averaging the redundant signals is beneficial, absent the dominant mean error arithmetic summation encountered in Figure 4-7. Improvement consequently corresponds to the per-sensor error divided by the square root of the number of combined sensors. Homogeneous sensor fusion with nonminimum mean error per sensor, however, clearly is not a viable signal conditioning methodology.

Heterogeneous multisensor data permits the integration of nonoverlapping information from independent sources for feature identification and data characterization unavailable from single sensors. Process automation systems are multivariable information structures that benefit from data attribution beyond that available at physical apparatus boundaries. The manufacture of contemporary aerospace composite materials illustrates a product performance example amenable to sensor-fusion-directed goals in conjunction with qualitative rule-based control. This schema overcomes limitations of conventional prescriptive

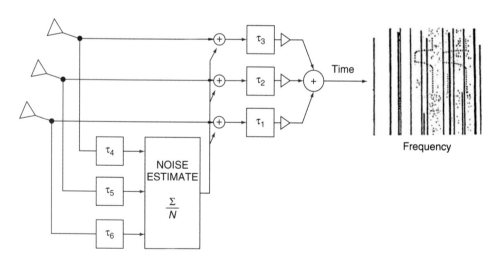

Figure 7-4. Sonar homogeneous sensor fusion.

autoclave composite cure with temperature and pressure controlled variables via augmentation by qualitative reasoning, diagrammed in Figure 7-5.

Embedded thermocouples measuring laminate in-situ temperature gradients are evaluated in concert with laminate-embedded dielectric impedance measurements of product cure gelation to preclude destructive overcure as a collaborative utility. Crucial end cure detection is signified by the sensor fusion decrease in dielectric impedance, evident only in the 100 Hz excitation frequency range, at laminate temperatures of 350°F or greater, shown in Figure 7-6. The error associated with this decision may be evaluated by the addition algorithm of Table 7-3, combining the mean and RSS errors of contributing instrumentation channels. It is notable that the performance of any control system cannot achieve less variability than the uncertainty associated with its end-to-end instrumentation error regardless of control sophistication.

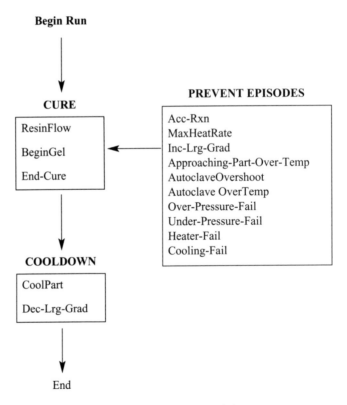

Figure 7-5. Autoclave cure rule base.

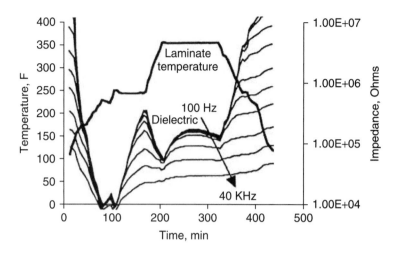

Figure 7-6. Autoclave process heterogeneous sensor fusion.

Interest in early warning systems for watershed monitoring employing online toxicity monitors has increased because of human health concerns from environmental contamination and vulnerability to intentional contamination events. In-situ watershed detection of toxic contamination is especially of interest because it enables immediate upstream water intake closure for remediation. However, electrochemical-based quantitative real-time measurement of all possible contaminants is so impractical that a search for alternative sensors has led to aquatic sentinel organisms that can express stress from exposure to toxic contaminants. Leading candidates are freshwater bivalve clams whose reduced gape expresses organism stress in response to broad-spectrum water-quality derogation by closing their shells to isolate vulnerable tissues. Thirty clams in a tray permit statistically valid samples measured by magnetic switches attached to each shell. Also, unattended monitoring is allowed because bivalves feed on naturally present water fleas and other organisms in the water flow.

Many watershed source intakes are remote, with no electric utility access as well as neither wired nor mobile communications services. Figure 7-7 describes a self-contained remote-telemetry early warning system, including heterogeneous sensor fusion data acquired from a bivalve biological sensor combined with an environmental YSI Model 6820 multisensor water sonde measuring temperature, pH, dissolved oxygen, conductance, and turbidity. A test of this solar powered 50-mile range 30 MHz HF system is shown, displaying upstream salt

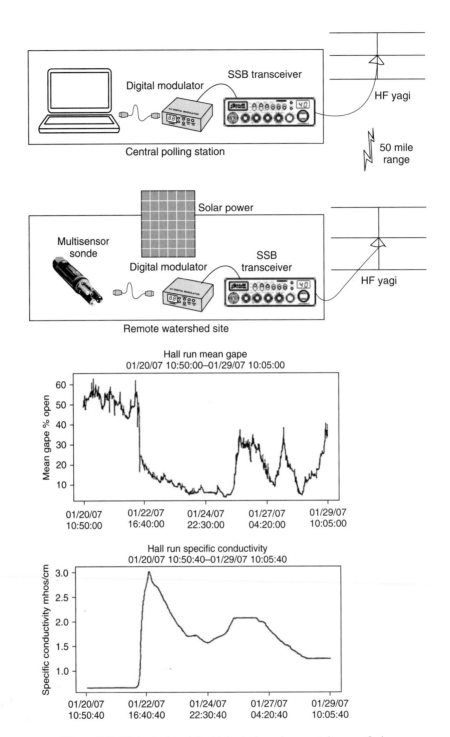

Figure 7-7. Watershed toxicity biological–environmental sensor fusion.

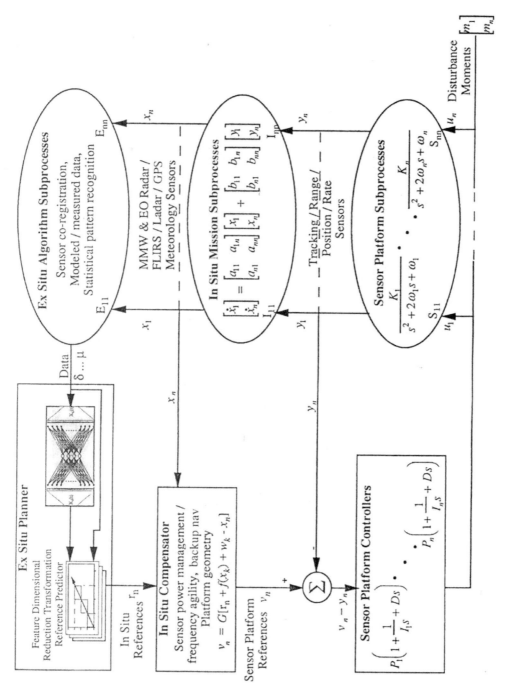

Figure 7-8. UAV avionics sensor fusion architecture.

191

additions over a one-week period. Notable is the synchronism of bivalve stress gape closure with saltwater conductivity increase. Figure 7-8 diagrams an unmanned aerial vehicle sensor fusion avionics architecture for integrating heterogeneous multisensor data into hypotheses for identifying most likely object states aided by sensed operating environments and identified uncertainties. These goals are achieved by combining sensor data correlation features with hypothesis evaluation, whereby processing assets are also managed for solution efficiency. An evolving merit of the associated hierarchical subprocess architecture is its versatility and applicability to broader process automation applications described in the following chapters. Practical elements include decision planners that provide feedforward control direction of ideal subprocess states while attenuating unmodeled process disturbances to enable more focused control and disorder reduction. Difficulties surmountable by these elements include achieving reliable process measurements otherwise consigned to ambiguous process data, and precisely controlled variables that otherwise may be too coupled to other process parameters.

7-4 INSTRUMENTATION INTEGRATION AND INTERFACES

Technical evolution and economic influences have combined to define the integration of multisensor instrumentation systems. Electrical measurements have evolved for nearly two centuries since the invention of the galvanometer in 1820. Four distinct instrumentation categories exist for meeting contemporary measur- and excitation and measurement requirements, as described by the taxonomy illustrated in Figure 7-9. These include dedicated discrete measurement systems, remote sensor telemetry, multifunction virtual instruments, and analytical instrumentation for process feature acquisition. The description of these categories is introduced in this section, including bus and network interfaces.

In practice , the diversity of existing bus structures enables a delineation of capabilities among the four instrumentation categories. Figure 7-10 introduces basic computer-bus interfaces. Level-0 traces define intercomponent board connections characterized by signals specific to linked digital devices. Level-1 dedicated buses, such as the ISA bus, provide buffered subsystem peripheral component interfacing, including protocols to accommodate signal propagation delays. Level-2 system buses , such as the peripheral component interconnect (PCI) structure, offer comprehensive bus master services, including arbitration and concurrent operation management. Level-3 parallel buses enable peripheral extensions for Level-1 buses, including GPIB and small computer systems interface

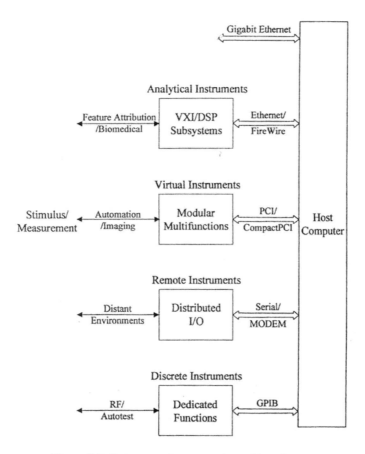

Figure 7-9. Instrumentation categories and interfaces.

(SCSI) buses. Level-4 serial buses are the longest structures in the bus repertory, and range from early standards such as RS-232C to the universal serial bus (USB). Level-5 video buses are typically limited to AGP display ports.

The general-purpose interface bus (GPIB) by Hewlett-Packard is a robust structure for networking discrete instruments in high-interference environments, including microwave frequencies. This parallel bus can link 15 instruments plus its controller with 16 active lines, eight for data and eight for control, as shown in Figure 7-11. Communications control procedures are initiated prior to data exchange designating transmitting and receiving ports. Instead of address lines, there are three data-transfer and five bus-management utility lines. In operation, when ATO is high all instruments must listen to the DIO lines, and when ATN is

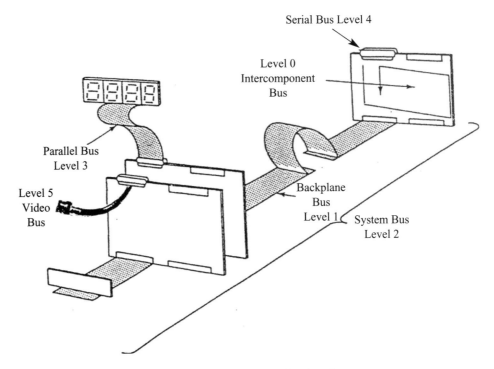

Figure 7-10. Basic computer-bus interfaces.

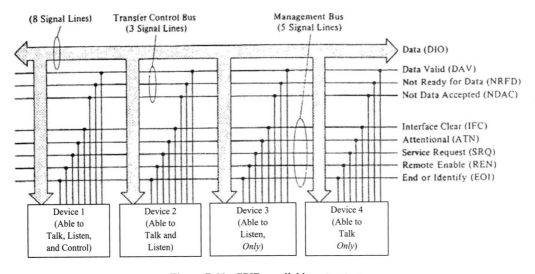

Figure 7-11. GPIB parallel bus structure.

low only designated instruments can communicate. External data exchanges with a host computer for all instrumentation categories is aided by the gigabit Ethernet, where performance is enhanced by PAM encoding, trellis error correction, and DSP-implemented signal equalization.

Computer-based automatic test equipment (ATE) has evolved as sophisticated implementations of parallel buses that link discrete instruments for the systematic validation of complex electronic systems. These systems employ versions of the IEEE Software Standard Abbreviated Test Language for All Systems (ATLAS), plus test executives in common scripted languages, including C++. These systems are commonly applied to military aircraft for plug-in flightline electronic system evaluation. By way of comparison, the remote sensing instrumentation category is more diverse, with applications ranging from distant environmental monitoring, introduced in Figure 7-7, to the satellite radiometer of Figure 1-24 in Chapter 1. In addition, efficient wideband peer-to-peer interfacing to 400 Mbps is available with the IEEE 1394 FireWire serial bus. FireWire interfacing is shown in Figure 7-9; however, interconnection distances are limited to 15 feet.

The peripheral component interconnect (PCI) is a versatile processor-independent computer bus structure illustrated in Figure 7-12, introduced by Intel Corporation for transfers of 64-bit words at up to 66 MHz rates, or 4 gigabits per second. None of the participating bus devices have dedicated memory or address assignments, but instead are configured by BIOS flash memory on power-up. Power-conserving reflected-wave logic switching is employed that requires only one-half logic-level voltage excitation. Bus bridge extenders are employed between separate PCI bus segments and to other buses such as ISA, permitting concurrent bus operations. Up to 256 PCI buses can be supported with the addition of bridges, each with a maximum of 256 peripheral devices. Compact PCI is an industrially hardened modular PCI bus available in a 3U or 6U Eurocard form factor for robust embedded applications. Applications include communications servers, industrial electronics, and defense electronic systems. The PCI bus bandwidth of 132 Mbytes per second also supports video data manipulation, whereas the PCI bus bandwidth of 8 Mbytes per second cannot.

Serial baseband signaling provides the majority of peripheral device and instrumentation system connections to host computers. Local area networks (LANs) have distinct functionalities. For example, computer LANs integrate network access devices with hosts and servers, including universal asynchronous receiver and transmitter (UART) terminal devices. This structure is described in Figure 7-13a. Source encoding commonly uses the RS-232C standard, shown as a full-duplex null-MODEM connection in Figure 7-13b that is capable of data rates to 600 bps at distances of 1 mile. Speed versus distance is principally determined

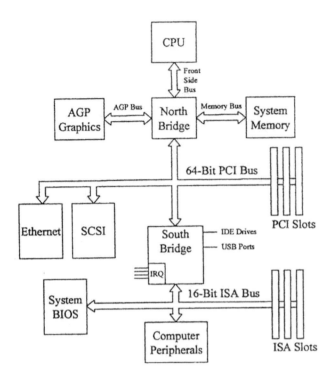

Figure 7-12. Exemplar PCI bus architecture.

by intersymbol interference with increasing distance. For noisy applications, RS-485 adds differential line drivers and receivers to the UART devices, whose common-mode interference rejection permits a data rate to 1 Kbps at distances of 1 mile, with up to 32 addressable communication nodes per port. This connection is shown in Figure 7-14. The higher performance USB port permits consolidation of peripheral interfacing by supporting up to 127 peripheral devices at data rates to 12 Mbps. This is a polled bus utilizing packet data exchange with automatic peripheral enumeration by its bus controller. However, USB hub-to-peripheral distances are limited to 15 feet.

Public LANs rely upon external network access such as Ethernet. Ethernet is a universal network currently employed worldwide with performance to 100 Mbps and connectivity allowable with twisted-pair as well as coax and fiber media. Its carrier-sensed multiple access, collision detection (CSMA/CD) datalink protocol benefits both from simplicity and effectiveness. Twisted-pair Ethernet (10 Base T) supports 10 Mbps, whereas fiber media Ethernet (100 Base FX) per-

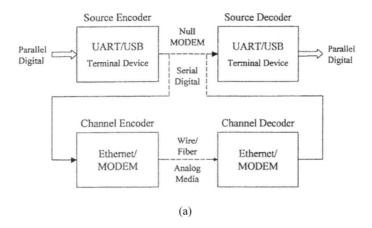

(a)

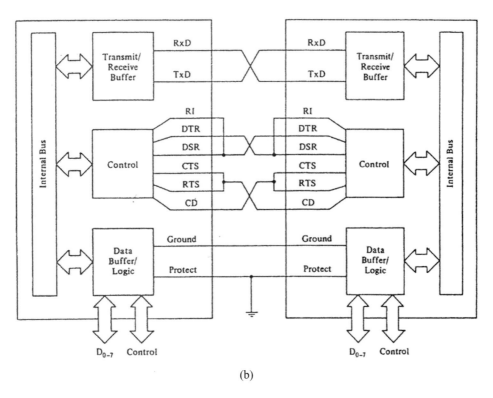

(b)

Figure 7-13. (a) Serial bus network. (b) RS-232C null-MODEM.

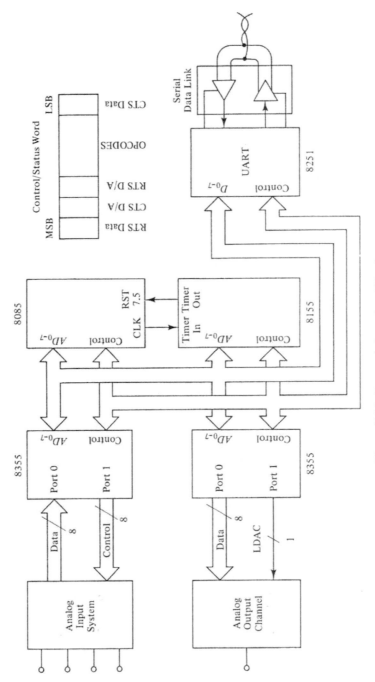

Figure 7-14. Remote independent I/O subsystem.

mits data rates to 100 Mbps. Ethernet employs a bus network bus topology and packet data format with a 48-bit unique worldwide address with allowable message size ranging from 512 bits to 1512 bytes. Gigabyte Ethernet (1000 Base T4) was introduced previously.

The expansion of virtual and analytical instrumentation systems, including the example of Figure 1-22 and process automation applications described in the following chapters, has led to specialized LANs for industrial applications. These are exemplified by Foundation Fieldbus and the controller area network (CAN). Fieldbus offers twisted-pair connectivity with a data rate of 31.25 Kbps to 1 mile, and permits remote devices to be powered over the same signal pair. CAN was intended for onboard automotive digital functions; however, its low- and high-speed data rate options, respectively 125 Kbps and 1 Mbps, has expanded its applicability since it supports distances to 1640 feet. Process instrumentation and control systems often benefit from node-to-node communication directly without interaction with a host computer. Such a local operating network (LON) protocol is offered by the Echelon Corporation as LonWorks. LonWorks employs serial packet data exchange over twisted pair at data rates of 78 Kbps to 4000 feet and 1.25 Mbps to 400 feet.

A remote independent I/O subsystem permitting tailored performance for local process elements is described by Figure 7-14. It incorporates onboard programmable computation to actuate data transactions with a shared FIFO buffer implemented with 8155 RAM. That is interactively coupled to a central host computer by means of the serial data link. This structure beneficially frees the host from a high overhead burden of generating comprehensive I/O timing and control signals. Remote independent I/O operation is activated by a local interrupt-driven service routine initiated by the onboard timer. A complete cycle acquires present output D/A data from the host through the buffer. Then A/D converts and stores, or transmits, input signals through the buffer. The control/status word shown provides synchronization between the host and remote I/O subsystem. When the host is free to accept new data it sets the CTS Data flag to initiate data transmission by the half-duplex 8255 UART-connected RS-485 serial data link.

PROBLEMS

A SETI biological space probe telemetry channel outputs a harmonic 1 V_{pp} analog signal with a dc-to-1 KHz BW uniform spectal occupancy that is to be digitally archived. Design two data conversion and recovery systems for this signal, providing a contrast in data conversion philosophies, excluding sensor and signal conditioning elements. For both A and B system block diagrams shown, provide comprehensive end-to-end one-sigma and six-sigma error budgets, including the components specified and system contributions showing all calculations. Recovered signal full-scale V_o is to be 10 V_{pp} for both systems.

7-1. System A (see Figure 7-15). Active filter output function interpolation with f_s/BW of 50 to minimize both noise aliasing and sinc attenuation. Employ AD779 and AD7840 data converters and OP177 amplifiers.

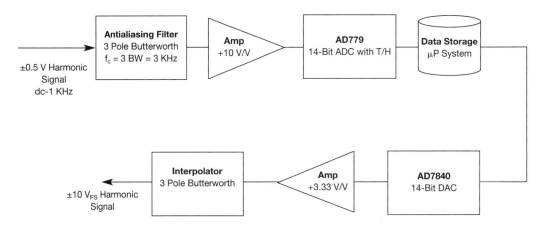

Figure 7-15. Design problem A. Space probe telemetry conversion/recovery system.

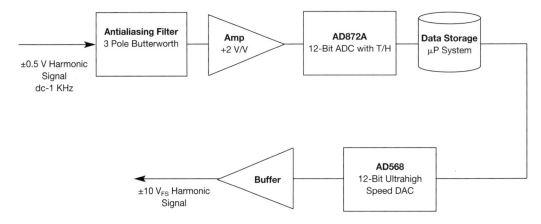

Figure 7-16. Design problem B. Space probe telemetry conversion/recovery system.

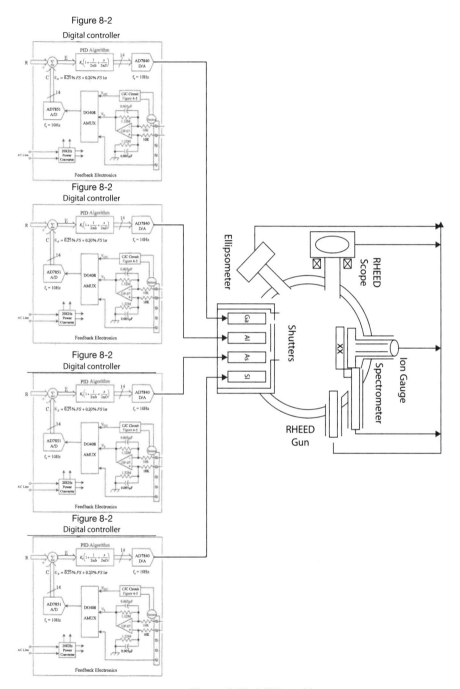

Figure 7-17. MBE machine.

7-2. System B (see Figure 7-16). Direct-D/A output interpolation with an f_s/BW of 5000. Employ AD872 and AD568 data converters and OP177 amplifiers.

7-3. The MBE machine (Figure 7-17) is employed for the production of thin-film multilayer optical and electronic devices. Production generally requires individual effusion-cell element-temperature stability sufficient to achieve on the order of 1% lattice-matched film growth for acceptable quality. Growth durations can extend between 1 and 5 hours, typically, as a function of cell-element flux magnitude, with up to three shutters open concurrently such as in the production of GaAlAs.

It is notable that the example thermal process controller parameters associated with Figure 8-2 are also descriptive of each closed-loop MBE effusion cell in this problem. Accordingly, identify the multisensor architecture represented by the composite of the cell control loops, and evaluate the combined temperature error of three cells simultaneously in their influence on film growth stability, applying Table 7-3 to calculate their error propagation.

BIBLIOGRAPHY

1. Abernethy, R.B. et al., "Uncertainty in Gas Turbine Measurements." Pratt & Whitney and Arnold Engineering Develop. Ctr, AEDC-TR 73-5, February 1973.
2. Brockman, J.P., *Error Referenced Non-Redundant Sample Coding For Data Conversion Systems,* M.S. Thesis, Elect. and Comp. Engr., Univ. of Cincinnati, 1986.
3. Dorsey, P.A., *Watershed Sensor Network Non-Line-Of-Sight Data Telemetry System,* M.S. Thesis, Elect. and Comp. Engr., Univ. of Cincinati, 2007.
4. Garrett, P.H., "Advances in Low-Frequency Radio Navigation Methods," *IEEE Trans. Aerospace and Electronic Systems,* Vol. AES-11, No. 4, July 1975, 562-574.
5. Garrett, P.H., "Vertical Path Guidance Computer," *IEEE Trans. Aerospace and Electronic Systems,* Vol. AES-6, No. 4, July 1970, 450–454.
6. Garrett, P.H., *High Performance Instrumentation and Automation,* CRC Press, 2005.
7. Gini, M., *Intelligent Autonomous Systems 7,* IOS Press, 2002.
8. Hall, D.L., *Mathematical Techniques In Multisensor Data Fusion,* Artech House, 1992.
9. Hill, B.K., *High Accuracy Airflow Measurement System,* M.S. Thesis, Elect. and Comp. Engr., Univ. of Cincinnati, 1990.
10. Lasorso, D.V., *Watershed Security Telemetry Network Protocol For Reliability Assurance,* M.S. Thesis, Elect. and Comp. Engr., Univ. of Cincinnati, 2009.
11. Peled, U., "A Design Method With Application To Prefilters And Sampling Rate Selection In Digital Flight Control Systems," Aeronautics Dept., Stanford University, 1978. Source: University Microfilms International, Ann Arbor, MI.
12. *Unmanned Aircraft Systems Roadmap 2005–2030 with Appendices.* Office of U.S. Secretary of Defense, Washington, D.C., 2005.

APPENDIX

LEMMA 1

Continuity equation for the reference plenum provides inlet V_{r1}:

$$\dot{m}_r = \rho A_{r_1} V_{r_1} = \rho A_{r_2} V_{r_2} \qquad \text{reference plenum}$$

$$\dot{m}_p = \rho A_{p_1} V_{p_1} = \rho A_{p_2} V_{p_2} \qquad \text{part plenum}$$

$$\therefore V_{r_1} = \frac{\dot{m}_r / \rho}{A_{r_1}} = \frac{25 \text{ ft}^3 / \text{min}}{0.1724 \text{ ft}^2} = 145 \text{ ft} / \text{min}$$

LEMMA 2

Adjusting $\dot{m}_p/\rho$ equalizes plenums stagnation P_0 for inlet V_{p1}/V_{r1}:

$$P_0 = P_{p_1} + \tfrac{1}{2}\rho\, V_{p_1}^2 \qquad \text{total part pressure}$$

$$P_0 = P_{r_1} + \tfrac{1}{2}\rho\, V_{r_1}^2 \qquad \text{total reference pressure}$$

$$\therefore P_{p_1} - P_{r_1} + \tfrac{1}{2}\rho(V_{p_1}^2 - V_{r_1}^2) = 0 \text{ for } P_0 \text{ equalized}$$

$$\frac{V_{p_1}}{V_{r_1}} = \sqrt{1 - \frac{2(P_{p_1} - P_{r_1})}{\rho V_{p_1}^2}} \qquad \text{velocity ratio of plenums}$$

LEMMA 3

Bernoulli equation determines part exit V_{p2}:

$$P_0 = P_{p2} + \tfrac{1}{2}\rho\, V_{p2}^2 \qquad P_0 \text{ constant throughout plenum}$$

$$V_{p2} = \sqrt{\frac{2(P_0 - P_{atm})}{\rho}} \qquad \text{where part } P_{p2} = P_{atm}$$

LEMMA 4

Exit V_{p_2}/V_{r_2} and continuity equations offer identity for $\dot{m}_p/\dot{m}_r$:

$$\frac{V_{p_2}}{V_{r_2}} = \frac{V_{p_1}}{V_{r_1}} \cdot \frac{A_{p_1}}{A_{p_2}} \cdot \frac{A_{r_2}}{A_{r_1}}$$

$$= \frac{\dot{m}_p/\rho}{\dot{m}_r/\rho} \cdot \frac{A_{r_2}}{A_{p_2}} \qquad \text{continuity equation identity}$$

$$\therefore \frac{\dot{m}_p/\rho}{\dot{m}_r/\rho} = \frac{A_{p_1}}{A_{r_1}} \cdot \sqrt{1 - \frac{2(P_{p_1} - P_{r_1})}{\rho V_{r_1}^2}}$$

LEMMA 5

Continuity equation defines part flow area A_{p_2} rationalized by blower ratio:

$$A_{p_2} = \frac{\dot{m}_p}{\rho} \cdot \frac{1}{V_{p_2}} \cdot \frac{\dot{m}_r/\rho}{\dot{m}_r/\rho}$$

$$= \frac{\dot{m}_p}{\rho} \cdot \frac{1}{\sqrt{\dfrac{2[(\frac{1}{2}\rho V_{r_1}^2 + P_{r_1}) - P_{\text{atm}}]}{\rho}}} \cdot \frac{\dot{m}_r/\rho}{\dot{m}_r/\rho}$$

$$= \frac{\dot{m}_p}{\dot{m}_r} \cdot \frac{\dot{m}_r}{\rho} \cdot \frac{1}{\sqrt{V_{r_1}^2 \dfrac{2(P_{r_1} - P_{\text{atm}})}{\rho}}}$$

LEMMA 6

Air-density-independent part flow area:

$$A_{p_2} = \frac{A_{p1}}{A_{r_1}} \cdot \sqrt{1 - \frac{2(P_{p_1} - P_{r_1})}{\rho V_{r_1}^2}} \cdot \frac{\dot{m}_r}{\rho} \cdot \sqrt{V_{r_1}^2 + \frac{2(P_{r_1} - P_{atm})}{\rho}}$$

$$= \frac{A_{p1}}{A_{r_1}} \cdot \frac{\dot{m}_r / \rho}{V_{r_1}} \cdot \sqrt{\frac{\rho \cdot \dfrac{2(P_{p_1} - P_{r_1})}{V_{r_1}^2}}{\rho + \dfrac{2(P_{r_1} - P_{atm})}{V_{r_1}^2}}}$$

$$= A_{p1} \cdot \sqrt{\frac{\rho - 2(P_{p_1} - P_{r_1}) / V_{r_1}^2}{\rho + 2(P_{r_1} - P_{atm}) / V_{r_1}^2}} \qquad \text{part flow area proof}$$

LEMMA 7

Open part plenum limit proof for A_{p2}:

$$A_{p_2} = A_{p1} \cdot \sqrt{\frac{\rho - 2(P_{atm} - P_{r_1}) / V_{r_1}^2}{\rho + 2(P_{r_1} - P_{atm}) / V_{r_1}^2}} \qquad P_{p_1} = P_{atm}$$

$$= A_{p1} \cdot \sqrt{\frac{\rho + 2(P_{r_1} - P_{atm}) / V_{r_1}^2}{\rho + 2(P_{r_1} - P_{atm}) / V_{r_1}^2}}$$

$$= A_{p1}$$

CHAPTER 8

Instrumented Processes Decision and Control

8-0 INTRODUCTION

International competitiveness has expanded emphasis on the integration of product design with process systems whose complexity is increasingly challenging. The present chapter addresses a continuing interest in instrumented process systems integrated with computer-centered decision and control capabilities for achieving extended performance. Figure 8-1 illustrates a road map representing process design employing features increasingly measurable in real time by innovative sensor systems.

Process automation systems have a heritage of intellectual depth and practical development. Contemporary systems are introduced employing feedback regulation of process apparatus focused on designs that minimize process variability. These typically adopt proportional-integral-derivative (PID) controllers for their robustness and communications utilities. A conventional model-reference process control case study is evolved to a more versatile three-level decision-capable subprocess influence-control structure initially applied to aerospace manufacturing [11]. This is characterized by an upper-level feedforward planner that outputs references to intermediate-level in-situ controllers distinct from base-level process physical apparatus regulation. Beneficially, process planner execution minimizes long-time-constant disturbance singularities while in-situ control attenuates short-time-constant processing disorder.

A final section introduces empirical-to-intelligent process decision examples highlighted by a UV excimer laser ablation process for the manufacture of aero-

Advanced Instrumentation and Computer I/O Design, Second Edition. By Patrick H. Garrett
Copyright © 2013 the Institute of Electrical and Electronics Engineers, Inc.

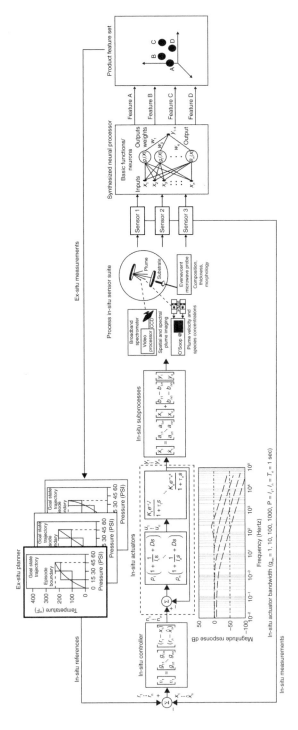

Figure 8-1. Instrumented processes decision and control road map.

208

space lubricants. Notably, fuzzy logic modeling of operator-domain knowledge is employed for the laser controller instead of modeling of the process itself.

8-1   PROCESS APPARATUS CONTROLLER VARIABILITY AND TUNING

Contemporary automation systems retain the requirement for feedback regulation of parameters within the process apparatus because process parameters and output are physical. Apparatus regulation is now integrated as the first level of a hierarchical subprocess architecture, as shown in the following section. The variability of process controllers, consisting of discrete instrumentation and automation components, may be quantified by measured and modeled device specifications and design choices from which these systems are fabricated. The resulting apparatus error analysis forecasts a baseline variability for process parameters, providing a measure of control influence on ultimate product quality. In these realizations, a majority of system implementations adopt commercial PID controllers for their robustness, sensor signal conditioning, and communications utilities. By some estimates, industrial automation systems employ one-in-three PID controllers for their efficient acquisition of process data alone, absent closed-loop control.

Figure 8-2 shows a representative Eurotherm Model 825 PID controller. The thermocouple signal conditioning analysis of Figure 4-5 is employed for the controller electronics front end to acquire the sensed process temperature variable T. The Type-C thermocouple is followed by a 22 Hz one-pole RC presampling low-pass filter and a differential subtractor instrumentation amplifier having a numerical voltage gain of 132. That raises a 31 mV dc full-scale input signal at 1800°C to 4.096 V for the data convertor.

The transfer function parameters of Equation (8-1) are for a dominant-pole process also included in Figure 8-2. When the process time constant τ_0 is known, Equation (8-2) may be employed to evaluate the analytically significant process closed-loop bandwidth –3dB response. The closed-loop bandwidth may also be evaluated experimentally from Equation (8-3) by plotting the controlled variable C rise time resulting from setpoint step excitation changes for R.

$$\frac{C}{R} = \frac{K_P K_C \left(1 + \dfrac{1}{2\pi I s} + \dfrac{s}{2\pi D}\right)}{1 + K_P K_C \left(1 + \dfrac{1}{2\pi I s} + \dfrac{s}{2\pi D}\right)} \cdot \left[\frac{\tau_0 s}{1 + K_P K_C \left(1 + \dfrac{1}{2\pi I s} + \dfrac{s}{2\pi D}\right)}\right] \quad (8\text{-}1)$$

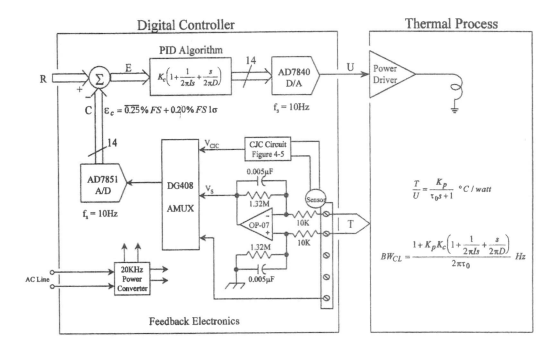

Digital Controller Instrumentation Error Summary

Element	$\varepsilon_{\%FS}$	Comment
Sensor	$\overline{0.011}$	Linearized thermocouple
Interface	$\overline{0.032}$	CJC sensor
Amplifier	0.103	OP-07A
Filter	$\overline{0.100}$	Signal conditioning
Signal quality	0.009	60 Hz ε_{coh}
Multiplexer	$\overline{0.011}$	Average tranfer error
A/D	0.020	14-bit successive approximation
D/A	0.016	14-bit actuation output
Noise aliasing	0.000049	–85 dB AMUX crosstalk from 40 mV @ 20 kHz
Sinc	$\overline{0.100}$	Average attenuation over BW_{CL}
Intersample	0.174	Interpolated by BW_{CL} from process τ_0
ε_c	$\overline{0.254}\%FS$	$\Sigma\overline{mean}$
	$0.204\%FS$	1σ RSS
	$0.458\%FS$	$\Sigma\overline{mean} + 1\sigma$ RSS

Figure 8-2. Process apparatus controller and variability.

$$BW_{CL} = \frac{1 + K_P K_C \left(1 + \dfrac{1}{2\pi I s} + \dfrac{s}{2\pi D}\right)}{2\pi \tau_0} \text{Hz} \quad \text{closed-loop bandwidth} \quad (8\text{-}2)$$

$$BW_{CL} = \frac{2.2}{2\pi t_r} \text{Hz} \quad \text{universal closed-loop bandwidth} \quad (8\text{-}3)$$

For simplicity of analysis, the product of combined controller, actuator, and process gains K is anticipated to approximate unity for a tuned control loop, and an example one-second process time constant enables the choice of an unconditionally stable controller sampling period T of 0.1 s by the developments of Figure 8-3. The denominator of the z-transformed transfer function defines the joint influence of K and T on its root solutions and, hence, stability within the z-plane unit circle stability boundary. Inverse transformation and evaluation by substitution of the controlled variable $c(n)$ in the time domain analytically reveals a 10–90% amplitude rise time value of 10 sampling periods, or one second, for unit-step excitation. Equation (8-3) approximates a closed-loop bandwidth value of 0.35 Hz. Table 8-1 provides definitions for symbols used in this example.

Examination of Figure 8-2 also reveals Analog Devices Corporation linear and digital conversion components with significant common-mode interference attenuation, shown with the input-signal conditioning amplifier of Table 4-3 and Figure 4-5. The corollary presence of 40 mV of 20 KHz power converter noise also results in negligible input crosstalk as coherent noise aliasing. A significant outcome is the closed-loop bandwidth influence on interpolation of the controller D/A output by attenuating its sampled-data image spectra. The resulting intersample error 0.174%FS of the controller output U is a dominant contribution to the instrumentation error summary. The 0.458%FS one-sigma total controller error approximates eight-bit binary accuracy. It is meaningful that this total controller error around the process apparatus loop defines the residual variability between the NIST-traceable true temperature and the controlled variable C measured temperature, including when C has achieved equality with the set point R. Note that this error cannot be reduced by skill in controller tuning.

Tuning methods are described in Figure 8-4 that insure stability and robustness to process disturbances by jointly involving process and controller dynamics online. Controller gain tuning adjustment outcomes generally result in a total loop

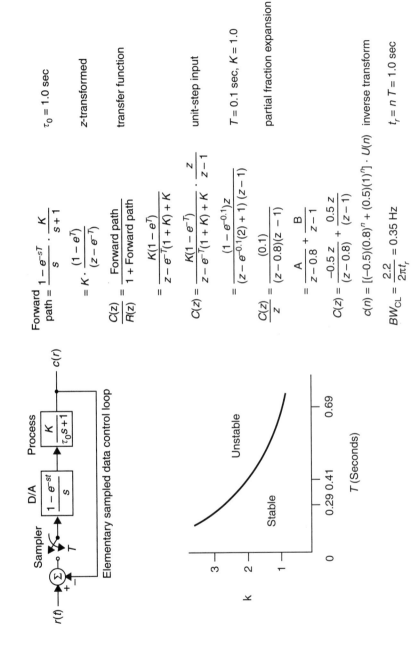

Forward path $= \dfrac{1-e^{-sT}}{s} \cdot \dfrac{K}{s+1}$ $\tau_0 = 1.0$ sec

$= K \cdot \dfrac{(1-e^T)}{(z-e^{-T})}$ z-transformed

$\dfrac{C(z)}{R(z)} = \dfrac{\text{Forward path}}{1+\text{Forward path}}$ transfer function

$= \dfrac{K(1-e^T)}{z-e^{-T}(1+K)+K}$

$C(z) = \dfrac{K(1-e^{-T})}{z-e^{-T}(1+K)+K} \cdot \dfrac{z}{z-1}$ unit-step input

$= \dfrac{(1-e^{-0.1})z}{(z-e^{-0.1}(2)+1)(z-1)}$ $T=0.1$ sec, $K=1.0$

$\dfrac{C(z)}{z} = \dfrac{(0.1)}{(z-0.8)(z-1)}$ partial fraction expansion

$= \dfrac{A}{z-0.8} + \dfrac{B}{z-1}$

$C(z) = \dfrac{-0.5\,z}{(z-0.8)} + \dfrac{0.5\,z}{(z-1)}$

$c(n) = [(-0.5)(0.8)^n + (0.5)(1)^n] \cdot U(n)$ inverse transform

$BW_{CL} = \dfrac{2.2}{2\pi t_r} = 0.35$ Hz $t_r = nT = 1.0$ sec

Figure 8-3. Feedback control closed-loop bandwidth.

Table 8-1. Process apparatus controller legend

Symbol	Dimension	Comment
R	°C	Controller setpoint input
C	°C	Process controlled variable
E	°C	Controller error signal
K_C	watts/°C	Controller proportional gain
I	sec	Controller integral time
D	sec	Controller derivative time
U	watts	Controller output actuation
s	rad/sec	Complex variable
K_P	°C/watts	Process gain
τ_0	sec	Process time constant
t_r	sec	Process response rise time
BW_{CL}	Hz	System closed-loop bandwidth
T	°C	Process sensed variable
V_{CJC}	mV/°C	Cold junction compensation
V_{OFS}	$4.096\ V_{pk}$	Full-scale process variable value
V_s	volts	Process variable signal value

gain of approximately unity when the process gain is included. The integrator value I provides increased gain near 0 Hz to obtain zero steady-state error for the controlled variable C. That effectively furnishes a control loop passband for accommodating the bandwidth of the error signal E. The lead derivative time D value enhances the controller transient response for both setpoint and process load changes to achieve reduced time required for C to equal R.

Sensor

Type-C thermocouple 17.2 mV/°C post-conditioning linearization

software $\dfrac{\overline{0.2°C}}{1800°C} \cdot 100\% = \overline{0.011}\%FS$

Interface

AD 590 temperature sensor cold-junction compensation

$\dfrac{\overline{0.5°C}}{1800°C} \cdot 100\% = \overline{0.032}\%FS$

Signal quality

$$\varepsilon_{\text{coh}} = \frac{V_{\text{cm}}}{V_{\text{diff}}} \cdot \left[\frac{R_{\text{diff}}}{R_{\text{cm}}} \right]^{1/2} \cdot \frac{A_{V_{\text{cm}}}}{A_{V_{\text{diff}}}} \cdot \left[1 + \left(\frac{f_{\text{coh}}}{f_c} \right)^{2n} \right]^{-1/2} \cdot 100\%$$

$$= \frac{(1 \ V_{\text{rms}} 2\sqrt{2})_{\text{pp}}}{31 \ \text{mV dc}} \cdot \left[\frac{80 \ \text{M}\Omega}{200 \ \text{G}\Omega} \right]^{1/2} \cdot \frac{0.02}{132} \cdot \left[1 + \left(\frac{60 \ \text{Hz}}{22 \ \text{Hz}} \right)^2 \right]^{-1/2} \cdot 100\%$$

$$= 0.009\% \text{FS}$$

$\varepsilon_{\text{ampl}_{\text{RTI}}}$	OP07A
V_{OS}	$\overline{10} \ \mu\text{V}$
$\dfrac{dV_{\text{OS}}}{dT} \cdot dT$	$2 \ \mu\text{V}$
$I_{\text{OS}} \cdot R_i$	$\overline{3} \ \mu\text{V}$
$V_{N_{\text{pp}}}$	$4.4 \ \mu\text{V}$
$f(A_V) \cdot \dfrac{V_{O\text{FS}}}{A_{V_{\text{diff}}}}$	$\overline{3} \ \mu\text{V}$
$\dfrac{dA_V}{dT} \cdot dT \cdot \dfrac{V_{O\text{FS}}}{A_{V_{\text{diff}}}}$	$15.5 \ \mu\text{V}$
$\Sigma\overline{\text{mean}} + \text{RSS}$	$(\overline{16} + 16) \ \mu\text{V}$
$X \dfrac{A_{V_{\text{diff}}}}{V_{O\text{FS}}} \cdot 100\%$	$0.103\% \text{FS}$

Analog multiplexer	
Transfer error	$\overline{0.01}\%$
Leakage	0.001
Crosstalk	0.00005
$\varepsilon_{\text{AMUX}}$	$\Sigma\overline{\text{mean}} + 1\sigma \ \text{RSS}$ $\overline{0.011}\% \text{FS}$

14-Bit A/D	
Mean integral nonlinearity (1 LSB)	0.006%
Noise + distortion (–80 dB)	0.010
Quantizing uncertainty ($\frac{1}{2}$ LSB)	0.003
Temperature coefficients ($\frac{1}{2}$ LSB)	0.003
$\varepsilon_{A/D}$ $\Sigma\overline{\text{mean}} + 1\sigma\,\text{RSS}$	0.020%FS

14-Bit D/A	
Mean integral nonlinearity (1 LSB)	0.006%
Noise + distortion (–80 dB)	0.010
Temperature coefficients ($\frac{1}{2}$ LSB)	0.003
$\varepsilon_{D/A}$ $\Sigma\overline{\text{mean}} + 1\sigma\,\text{RSS}$	0.016%FS

Noise aliasing

$$\varepsilon_{\text{coherent alias}} = \text{Interference} \cdot \text{AMUX crosstalk} \cdot \text{sinc} \cdot 100\%$$

$$= \frac{V_{\text{coh}}}{V_{O_{\text{FS}}}} \cdot -85\ \text{dB} \cdot \text{sinc} \cdot \left(\frac{mf_s - f_{\text{coh}}}{f_s}\right) \cdot 100\% \qquad m\ \text{defined at}\ f_{\text{coh}}$$

$$= \frac{40\ \text{mV}}{4096\ \text{mV}} \cdot (0.00005) \cdot \text{sinc}\left(\frac{2000 \cdot 10\ \text{Hz} - 20\ \text{Hz}}{10\ \text{Hz}}\right) \cdot 100\%$$

$$= 0.000049\%\text{FS}$$

Sinc

$$\varepsilon_{\text{sinc}} = \frac{1}{2}\left(1 - \frac{\sin \pi BW_{\text{CL}} / f_s}{\pi BW_{\text{CL}} / f_s}\right) \cdot 100\%$$

$$= \frac{1}{2}\left(1 - \frac{\sin \pi\, 0.35\ \text{Hz} / 10\ \text{Hz}}{\pi\, 0.35\ \text{Hz} / 10\ \text{Hz}}\right) \cdot 100\%$$

$$= 0.100\%\text{FS}$$

Controlled variable interpolation

$$\varepsilon_{\Delta V} = \left[\frac{V_{O_{FS}}}{V_s^2 \cdot \left\{ \mathrm{sinc}^2\left(1 - \frac{BW_{CL}}{f_s}\right) \cdot \left[\left(1 + \frac{f_s - BW_{CL}}{BW_{CL}}\right)^2\right]^{-1} + \mathrm{sinc}^2\left(1 + \frac{BW_{CL}}{f_s}\right) \cdot \left[\left(1 + \frac{f_s + BW_{CL}}{BW_{CL}}\right)^2\right]^{-1} \right\}} \right]^{-1/2} \cdot 100\%$$

$$= \left[\frac{4.096V^2}{(4.069V)^2 \cdot \left\{ \left[\frac{\sin \pi\left(1 - \frac{0.35\,\mathrm{Hz}}{10\,\mathrm{Hz}}\right)}{\pi\left(1 - \frac{0.35\,\mathrm{Hz}}{10\,\mathrm{Hz}}\right)}\right]^2 \cdot \left[1 + \left(\frac{10\,\mathrm{Hz} - 0.35\,\mathrm{Hz}}{0.35\,\mathrm{Hz}}\right)^2\right]^{-1} + \left[\frac{\sin \pi\left(1 + \frac{0.35\,\mathrm{Hz}}{10\,\mathrm{Hz}}\right)}{\pi\left(1 + \frac{0.35\,\mathrm{Hz}}{10\,\mathrm{Hz}}\right)}\right]^2 \cdot \left[1 + \left(\frac{10\,\mathrm{Hz} + 0.35\,\mathrm{Hz}}{0.35\,\mathrm{Hz}}\right)^2\right]^{-1} \right\}} \right]^{-1/2} \cdot 100\%$$

$$= \left[\frac{1}{\left(\frac{0.001}{3.03}\right)^2 \cdot (0.001313) + \left(\frac{-0.1094}{3.251}\right)^2 \cdot (0.001142)} \right]^{-1/2} \cdot 100\%$$

$$= 0.174\%\,FS$$

Ziegler and Nichols quarter-decay PID tuning has been historically applied to control loops, with controller gain empirically determined from the achievement of a four-to-one subsidence ratio between the first two periodic peaks of a controlled variable response, following a set point step change. However, obtaining well performing integral and derivative terms by these approximations, described in Figure 8-4, are insufficient because of their cross coupling. That limitation motivated the development of the decoupled trapezoidal controller tuning method.

Trapezoidal tuning input power pulses provide trapezoidal controlled variable changes exhibiting a differing rate of rise to rate of fall from variations in process

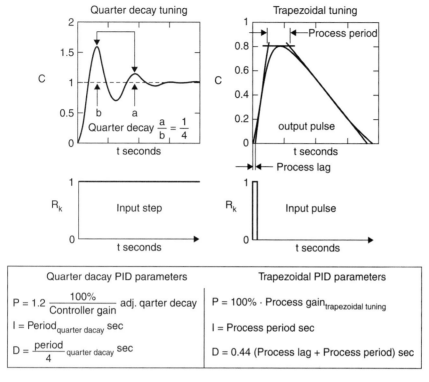

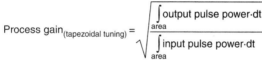

$$\text{Process gain}_{(\text{tapezoidal tuning})} = \sqrt{\frac{\int\limits_{\text{area}} \text{output pulse power} \cdot dt}{\int\limits_{\text{area}} \text{input pulse power} \cdot dt}}$$

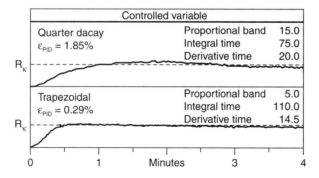

Figure 8-4. Quarter decay versus trapezoidal tuning.

dynamics between these two states, shown in Figure 8-4. Since the product of process and controller gains ideally approximates unity around a control loop, and process gain may be defined as the ratio of measured power output to power input over time intervals defining their respective baseline areas, then controller proportional band P is defined as the process gain times 100% or simply the inverse of controller gain.

Proportional band P is defined as the percentage of full-scale change in controller input signal required to move the controller output to its full-scale value. For example, a 10% proportional band will produce a full-scale controller output for a 10% controller input signal change, hence, a controller gain of 10 as $100\%/P$.

The controller integral period I is independently derived from the process experimental response period, to define the bandwidth in Hz of $1/2\pi I$ s for a controller error signal E, for minimizing the steady-state error of the controlled variable C. The expression for derivative time D is determined from the piece-wise-linear fit to the pulsed cell response to achieve a high-pass response beginning at $s/2\pi D$ in Hz. Example set point R step changes to the control loop of Figure 8-2 are recorded as the controlled-variable C responses in Figure 8-4. Results disclose a steady-state error on the order of 0.29% of setpoint for trapezoidal tuning that corresponds to the controller error summary of Figure 8-2, which is only one-sixth the 1.85% error for quarter-decay tuning.

Production processes require essential properties of composition and structure to be identified and translated into physically realizable apparatus parameters with sufficient resolution and accuracy to meet product goals. Figure 8-5 describes an example process apparatus for chemical vapor deposition of a lanthanum hexaluminate coating on embedded sapphire filaments for strengthening high-temperature ceramic matrix composites employed in aerospace vehicles and aircraft engine applications. Coating recipe properties providing a gradient between the ceramic composite and fiber filament morphologies impart critical functional strength to the final material.

Process apparatus actuators include 25 control valves, eight mass-flow-gas regulators, 13 PID controllers, and one each fiber velocity, precursor high-pressure liquid pump, and compressed-air purge-flow control. Liquid precursors permit precise regulation of coating stoichiometry, and reactor gas stream oxygen regulation provides accurate coating temperature setpoints, enabling continuous reliability in the production of coating properties. Mass spectrometry in concert with Raman spectrometry allow, respectively, comparison of in-process wet chemistry and postprocess product chemistry. Analysis of this example continues with Figure 8-12 following subsequent process control evolution.

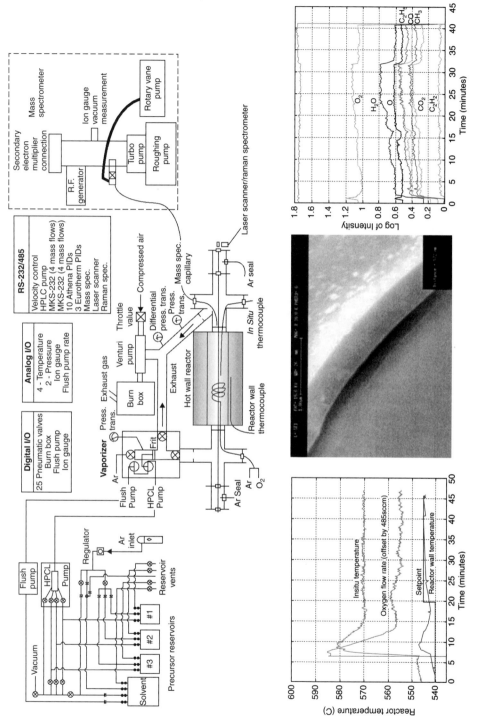

Figure 8-5. Chemical vapor deposition process apparatus.

8-2 MODEL REFERENCE TO REMODELING ADAPTIVE CONTROL

PID controllers employed for apparatus regulation also impart necessary process stabilization and variability minimization, but alone are incapable of meeting complex control requirements in advanced laboratory and industrial process systems. Specifically, critical processing states may require extended process description and control capabilities in order to achieve essential processing goals. In addition to regulation of process mass, momentum, and energy model-reference control extends that capability to interact directly with physical and chemical quantities within a process system. This is illustrated by a quantitative model for selecting between scrubber reagent addition controllers.

In Figure 8-6 a U.S. EPA prototype limestone reagent sulfur dioxide absorber is sized for a common 100 megawatt (134,000 horsepower) coal-fueled utility boiler. This scrubber is scaled to recirculate 16,000 liters of limestone slurry per thousand cubic meters of combustion flue gas to meet a goal of 90% sulfur removal. Limestone reagent offers the advantage of one-tenth the cost of lime, but produces periodic process disturbances when scrubber inlet flue gas load decreas-

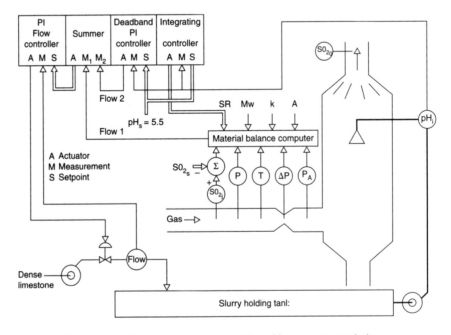

Figure 8-6. Stoichiometric-assisted pH scrubber reagent regulation.

es are encountered. That specifically results in elevated slurry pH, which is exacerbated by normally extended recirculating limestone dissolution times. The net process influence of this behavior is graphed by the declining scrubber reaction gain of Figure 8-7. Further, minimizing internal scrubber calcium scaling while maintaining consistent sulfur dioxide removal imposes a constrained narrow range of slurry pH centered about a 5.5 value.

These processing complexities can benefit from the control of limestone feed rate based on the stoichiometric ratio of entrained scrubber calcium-to-sulfur dioxide obtained from measurements of slurry pH and inlet flue gas constituents. Automatic selection accordingly occurs between the two flow controllers, described in Figure 8-6, each independently scaled for full limestone addition as input to their common summer. A model-reference feedforward stoichiometric controller employs inlet gas constants and measurements to calculate a material-balanced limestone feed rate, described by Equations (8-4) and (8-5), where slurry Flow 1 relies on volumetric flue gas flow Q. An integrating memory controller accordingly adjusts the feedforward gain stoichiometric ratio (SR) value, decreasing SR for an increasing pH, in response to differences between slurry inlet pH and the scrubber 5.5 pH setpoint. SR is adjusted along the vertical axis of Figure 8-8, where the SR controller gain within the 5.3 to 5.7 pH dead

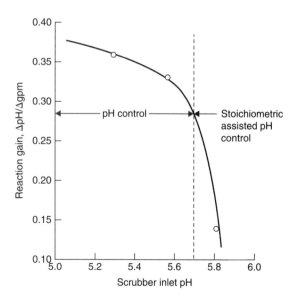

Figure 8-7. Scrubber reaction gain versus pH.

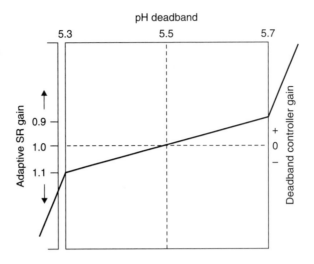

Figure 8-8. Stoichiometric and pH deadband gain.

band is low, and variable beyond to recapture SR control in the event of pH malfunction.

$$\text{Flow 1} = (\text{inlet variables})(\text{feedforward gain})(\text{scaling}) \tag{8-4}$$

$$= (\Delta SO_2 \times Q)(SR)\left(\frac{\text{moles Ca / moles} SO_2}{\text{limestone purity} \times \text{slurry lb / gal} \times \text{density} \times \text{specific gravity}} \right)$$

$$= (\Delta SO_2 \times Q)(SR)\left(\frac{100 / 64}{0.97 \times 8.3 \times 0.6 \times 1.61} \right)$$

$$= \Delta SO_2 \times Q \times SR \times 0.2 \qquad \text{gallons per minute}$$

$$Q = A \cdot k \sqrt{\frac{\Delta P \cdot T}{P \cdot M_w}} \cdot \frac{P \cdot T_{std}}{P_A \cdot T} \qquad \text{cubic meters / minute} \tag{8-5}$$

Stoichiometric-assisted pH regulation of limestone feed rate enhances scrubber performance, extending to low process reaction gains that pH-only feed rate regulation cannot achieve. Feedback pH regulation of limestone feed rate is based on the slurry aggregate of scrubber reactions, including sulfur dioxide neutralization, boiler flyash alkali scrubber entrainment, and chloride additions from coals, plus make-up water to replace the constant 300°F evaporation. Equation (8-6) defines pH controller limestone feed rate Flow 2, whose pH sensor span of 4.5 to 6.5 increases control responsiveness. Discrete modular components are utilized throughout the system implementation, including electrical isolation of the material-balance computer, to assure reliability in this high-energy operating environment. It is notable that the Flow 1 and Flow 2 reagent controllers are combined by their associated dead-band controller. That introduces an axiomatic principle of decoupling in multivariable control developed later in this section.

$$\text{Flow 2} = (\text{pH}_s - \text{pH}_i)(\text{controller gain})(\text{scaling}) \qquad (8\text{-}6)$$

$$= (\text{pH}_s - \text{pH}_i)(0.5)\left(\frac{8\,\text{gpm max}}{2.0\,\text{pH span}}\right)$$

$$= (\text{pH}_s - \text{pH}_i) \times 2.0 \qquad \text{gallons per minute}$$

Extension of process modeling permits detailed description of state information to comprehensively coordinate multiple, simultaneously controlled parameters, compensate migrating product features, and achieve near-ideal process-state progression. This is practically enacted by providing reference inputs to in-situ controllers that are directed by a feedforward modeled product planner. Performance further accrues from process decomposition into subprocesses to allow more precisely modeled, and potentially more linear, subprocess regions. The degree of decomposition depends upon the modeling granularity required and natural hierarchy presented.

Illustration of the required control engineering schema is shown in Figure 8-9, where product outputs are represented within the in-situ subprocesses whose in-situ actuators are realized by the feedback apparatus controllers and subprocesses shown in the dashed box. In the absence of in-situ feedback measurements to the in-situ controllers, however, in-situ subprocesses revert to extended apparatus subprocess lag elements. Product controllability is then limited to apparatus controller effectiveness in determining product goals, whereby any shortfall equates to elevated processing disorder. It is significant that with in-situ feedback, the cascade control relationship, specifically, apparatus time constants divided by in-situ controller gains $\tau_n/(1 + g_{nn})$, proportionally extend in-situ actuator closed-loop bandwidths and the frequency response of in-situ subprocesses. The B matrix coefficients of the in-situ subprocess models link the apparatus to in-situ subprocesses, and the A coefficients link the in-situ subprocesses to product subprocesses.

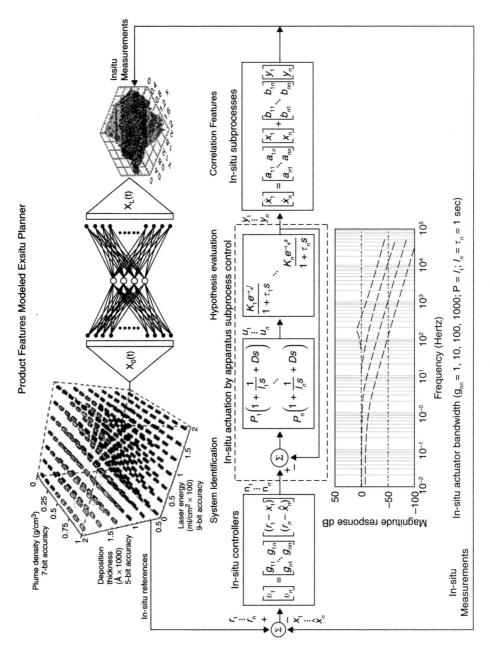

Figure 8-9. Product features remodeling adaptive control.

Many control systems benefit from the additional performance enabled by the decision capability of added online models of their process and products. When system identification of these decision-capable models also employ real-time process and/or product measurements and estimates for model updating, adaptive control is defined. An ex-situ planner implements a feedforward control algorithm [2, 10] approximating the inverse of the in-situ subprocesses. Planner elements include ideal product description episodes keyed to current process states, identified by real-time in-situ subprocess measurement inputs that output in-situ references r_n for control direction. These episodic in-situ reference outputs are modified by partially revised external product analytical assays, hence ex situ, aided by correlation for online hypothesis evaluation of potential product feature migration and planner episode remodeling. The ex-situ planner is thus updated by product structure and composition measurements, such as microwave microscopy by Figure 1-22, and qualitative feature estimates. Performance advancement includes feedforward planner attenuation of encountered long-time-constant product disturbances, and in-situ feedback minimization of short-time-constant subprocess disorder.

Initial understanding about process influences on product goals may be limited. It is therefore useful to document emerging understanding in terms of subprocess parameter linkages to assist improvement in process design, identify specific control requirements, and predict product outcomes. The hierarchical subprocess influence architecture of Figure 8-10 has subsequently evolved from the control engineering schema of Figure 8-9 to aid decomposition into three levels: apparatus regulation, adaptive in-situ controllers, and the product planner. A principal benefit of the influence diagram is its ability to reveal detailed subprocess parameter linkages and their influence on the final product. That information guides possible process modifications necessary to minimize intersubprocess parameter coupling for more robust product processing, often reducing required control complexity. Figure 8-11 mathematically portrays ideal matrix-diagonal term, uncoupled subprocess-to-subprocess direct linkages by solid lines, and process-disorder-inducing matrix-off-diagonal terms representing crosscoupled intersubprocess linkages by dashed lines. In practice, decoupling of intersubprocess parameters may be achieved through apparatus design or redesign that provide rectangular matrices [14] with as sparsely as feasible covariance terms, either upper or lower triangular, defining a compromise of decoupled intersubprocess linkages.

The hierarchical subprocess linkage influences of Figures 8-10 and 8-11 are demonstrated by their application to the chemical vapor deposition process apparatus of Figure 8-5. Figure 8-12 continues to describe specific materials subprocess linkage influences between apparatus, in-situ, and product levels of the

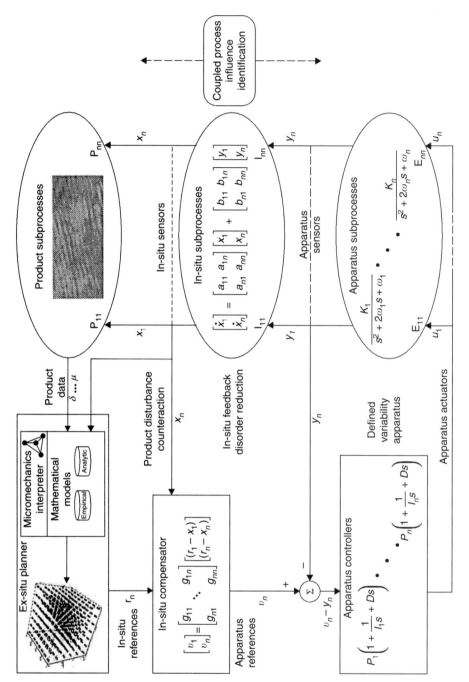

Figure 8-10. Hierarchical control subprocess influences.

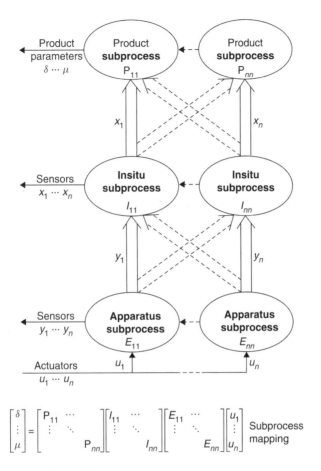

Figure 8-11. Axiomatic subprocess influences.

process hierarchy [8]. Essential fiber-coating-apparatus parameters are shown, including reactant temperature and reactant flow rate. A single uncoupled apparatus influence links fiber speed to in-situ coating growth rate, which corresponds to in-situ subprocess matrix diagonal term I_{11}. Apparatus reactor temperature has a direct influence on in-situ coating morphology, whose linkage is described by matrix term I_{22}. However, this temperature is also crosscoupled to both in-situ coating growth rate by I_{12} and chemical composition by I_{32}. The third apparatus parameter reactant flow rate is directly linked to in-situ chemical composition by I_{33}, but also crosscoupled to both in-situ coating morphology term I_{23} and growth rate term I_{13} in Figure 8-12.

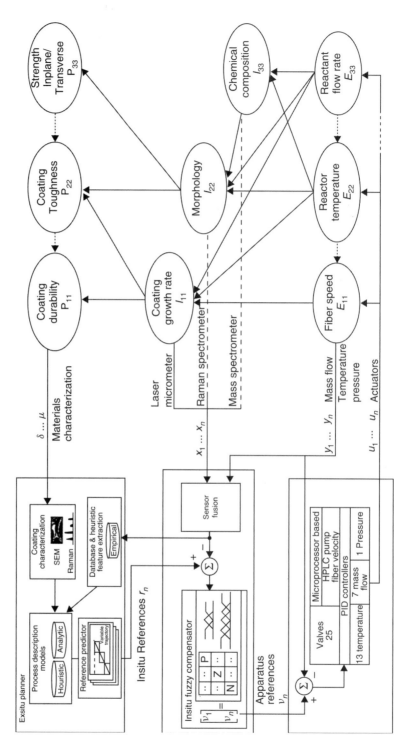

Figure 8-12. Chemical vapor deposition subprocess influences.

228

In-situ subprocesses growth rate and morphology are also linked to upper product subprocesses, respectively coating durability and toughness, by direct uncoupled intersubprocess matrix diagonal terms P_{11} and P_{22}. Also, decoupled in-situ growth rate and morphology linkages to product subprocesses, respectively coating toughness and strength, by matrix terms P_{21} and P_{32}. All actuator linkages are uncoupled, defining the matrix trace E_{11}, E_{22}, and E_{33}. The multiplication of these linkage terms, when substituted in the subprocess mapping matrices of Figure 8-11, define the combined influences on product characterization data as well as the planner models of Figure 8-9. An extended case study of this product description method is presented in Chapter 9 for superconductor manufacturing.

In-situ subprocess control of coating chemical composition is, therefore, the least influenced by apparatus parameter values with two linkages; a combination of reactant temperature and flow rate parameter values. In-situ control of coating growth rate is the second least influenced by apparatus parameter values; growth rate goal values may be met by a combination of apparatus reactant temperature and flow rate parameter values plus independently adjustable fiber speed. In-situ control of coating morphology emerges as the most influenced by apparatus parameter values; morphology goal values may be met only by an iterative combination of apparatus reactant temperature and flow rate parameter values, plus the dependent chemical composition goal value.

These self-documenting hierarchical subprocess parameters are highly informative to process design and control system realizations [12]. With available subprocesses identified in Figure 8-12, elective product recipes and subprocess parameter values may be implemented, aided by factorial experiments to determine specific values. The regulation of apparatus reactor temperature and reactant flow rate are implemented by fuzzy logic controllers because of their wideband closed-loop response and utility in minimizing process disorder. Fuzzy logic control design is developed in Section 8-3. Figure 8-5 also shows reactor temperature measurement by an in-situ gas-stream thermocouple with temperature actuation, supplying oxygen mass flow to the combustible gas. This permits reactor temperature to be regulated within 1°C of elected setpoints while wall temperature is fixed.

8-3   EMPIRICAL TO INTELLIGENT PROCESS DECISION SYSTEMS

Process controllability is concerned with the capability for achieving process goal states, where controlled variables are conventionally fewer than the number of process states. On the other hand, observability is concerned with the capability for estimating the values of process states, often through direct measurement of controlled variables. It follows that process controllability through observability fundamentally may govern control system outcomes. Further, progress in sensor materials, micromachining, and microelectronics technologies have advanced observability capabilities. In this context, the following example illustrates controllability through observability for an instrumented process empirical decision application.

A chemical chromate coating is commonly applied to galvanized steel strip during production for wide use in vehicle and construction products. This coating provides both a nonreactive barrier to environmental exposure to retard white rust formation, and a zinc restoration reaction following physical damage. Liquid chromate is sprayed onto the top and bottom of a moving 50–500 fpm horizontal galvanized strip, formulated as 6% chromate by volume, whose preponderance of water also functions as a rinse whose runoff is recirculated. Excess spray is removed with squeegees actuated by east and west pressure regulators acting at each squeegee roller end, where redundant chromate application and squeegee apparatus is in standby to enable continuous operation when failures occur. A uniformly controlled strip wet-film coating of 2.3 ml/m^2 is required at an exit gas-fired dryer to preserve chromate deposition uniformity at the goal value of 16.5 mg/m^2.

A pictorial of this process apparatus is provided by Figure 8-13. The control system design is shown in Figure 8-14 [3] and employs separate ratio PID controllers for chromate solution volumetric flow spray application to the strip top and bottom. These separately regulate the different application effects acting on the top and bottom spray processes through adjustment of 2–20 gph flow setpoints, including a function of the 50–500 fpm strip speed. The spray apparatus includes Eurotherm 900 EPC ratio controllers, Milton-Roy volumetric liquid pumps, and Hoffer turbine flow meters. Sensed process disturbances of top and bottom strip chromate solution nonuniformity are provided by separate scanning near-infrared reflectance sensors (Moisture Systems Corporation MQ series, introduced in Figure 1-21), providing wet-film thickness measurements in ml/m^2. Decoupled feedback control of east and west squeegee pressures, employing additional ratio PID controllers, effectively maintain the final chromate solution wet-film uniformity at 2.3 ml/m^2.

Cognition of and reasoning about the physical world is central to both human and machine intelligence, where physics and mathematics offer expressions for useful models and constraints to aid in this reasoning. Qualitative reasoning about decisions for physical systems is especially useful when quantitative models are

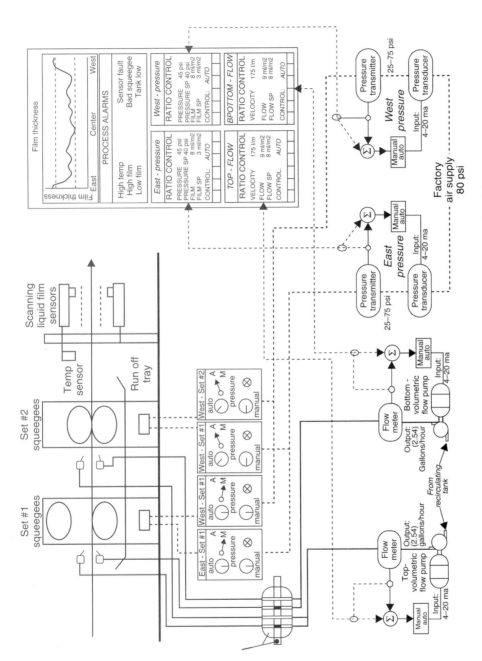

Figure 8-13. Chromate application process apparatus pictorial.

231

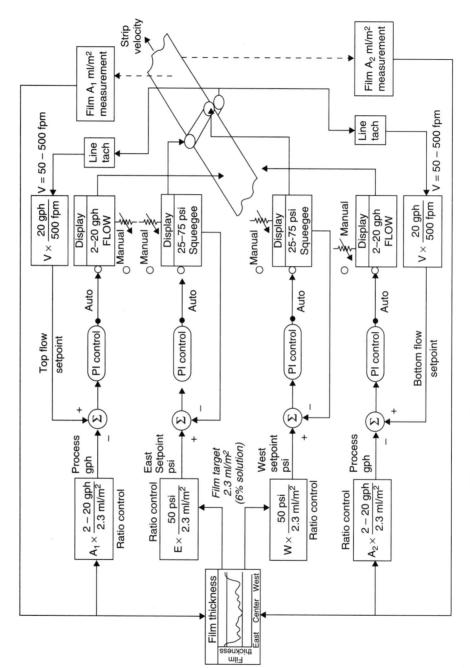

Figure 8-14. Chromate application empirical ratio control.

incomplete or unavailable. For example, computationally intelligent models may be implemented to achieve effective process direction in terms of IF, THEN, and ELSE decisions. They also aid recovery from extraordinary states such as process limit cycling, in which quantitative models typically are constrained to between 0 and 100% of a parameter range.

For process applications, a qualitative planner may provide state graphs describing process progression with a qualitative calculus that implements four tasks: determining executable process states, arbitrating appropriate state influences, resolving state conflicts, and translating quantitative-to-symbolic sensor and actuator values. A rule-based model usefully enables process execution as a function of goals and current events instead of a prescriptive state progression with no capability for changing pathways. Rule-based reasoning may therefore be embodied in feedforward process planner decision models for outputting ideal process state trajectories. Objects describing this planner are illustrated in Figure 8-15.

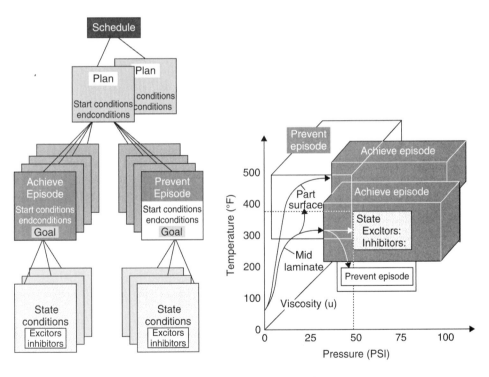

Figure 8-15. Rule-based process planner.

A qualitative process planner directs the aerospace composite process apparatus of Figure 8-16, whose actuation is provided by the temperature and pressure PID feedback controllers shown. Internal laminate product cure properties are sensed online during processing, as described in the sensor fusion example of Figure 7-6, to furnish current process state information for updating the rule-based planner model to aid decision execution. Plans contain episodes that define

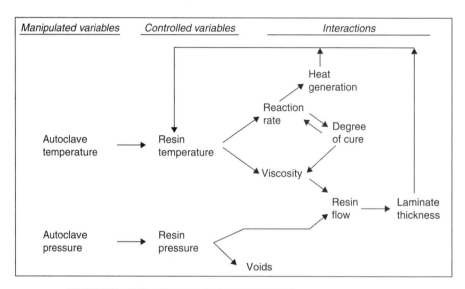

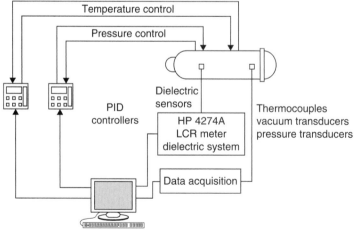

Figure 8-16. Autoclave apparatus parameter influences.

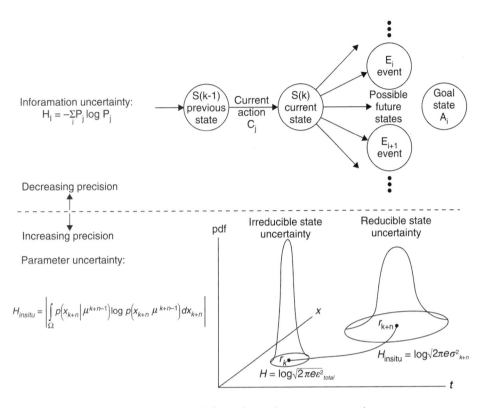

Figure 8-17. Process information and parameter uncertainty.

process parameters over a cure cycle, each of which describe goal states to be achieved, maintained, or prevented. The main cure plan has three episodes: pre-cure Resin Flow, limited to a 250°F autoclave temperature; Begin Gel, increasing temperature to 350°F to initiate curing; and End Cure, invoking cooldown following sensor fusion detection of viscosity parameters explained in Figure 7-5. Embedded laminate in-situ thermocouple temperature and dielectric resin viscosity sensors provide the sensor fusion data.

Process planner models benefit from efficient information representation measurable by an entropy criterion, ranging from $0 < H < 1$, providing a unified expression of uncertainty in process systems. Entropy satisfies the additive property whereby a combination of subsystems will be optimal when their total entropy is minimized, observing that at unity complete disorder is exhibited. Knowledge pooling consequently is useful for efficient process systems, in which entropy minimization corresponds to knowledge focusing and as few process actuations as practicable. Figure 8-17 illustrates control actions C, describing actuations toward processing goals A, with processing states S encountering possible disorder events E.

Modeled device and system errors defined in Figure 8-2 quantify closed-loop controller errors that forecast the uncertainty of per-control-loop apparatus variability. This is expressed as the irreducible entropy H in Figure 8-17, whose value can only increase to $H_{\text{in situ}}$ as a consequence of processing progression through disorder encounters. Apparatus variability and in-situ subprocess disorder were introduced in prior sections of this chapter including methods for their attenuation. Overarching these parameter uncertainty sources, however, is the potential information entropy arising from invalid rule selection. Figure 8-18 graphs how this uncertainty may be quantified by rule acceptance testing for a not-to-exceed number of disconfirms.

A structure for qualitative reasoning in intelligent control is now presented. Response to the need for robust controllers capable of accommodating nonlinear and time-migrating process models prompted the development of adaptive control concepts. However, since adaptive controllers may react to nonlinear process

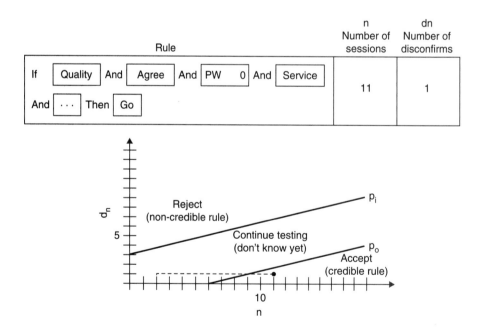

Figure 8-18. Expert rule confirmation.

conditions they also must become nonlinear entities whose realization has been difficult, but have benefitted from the evolution of computational intelligence. Intelligent control has evolved from expert systems, fuzzy logic, neural networks, and genetic algorithms. The following control case study employs fuzzy logic encoding of operator process domain knowledge with a symbolic actuation representation providing smoothing across the control domain. Fuzzy logic is a method that aids feedback control to mimic the performance of human cognition. Nevertheless, it can effectively approximate rigorously expressed parameters, whose uncertainty and variability are high, with rule-based formalisms. The objective of fuzzy control, therefore, is to model the behavior of operator-domain knowledge about a process instead of modeling the process itself. In effect, the fuzzy controller becomes a logical model of operator-domain knowledge for actuation of process parameters.

Elements of a fuzzy set can be linguistic variables; their membership degree will assume values in the range of 0 to 1, where 1 represents certainty. This formalism provides the dimension necessary to represent multivalued information. Fuzzy logic accordingly extends and adapts mathematical language to accommodate processes such as human reasoning rather than being constrained to bivalued computer logic. Fuzzy logic may also accommodate logic operations by utilizing IF–THEN statements. For example, IF water is hot, THEN cold water should be INCREASED. HOT and INCREASED are fuzzy sets. HOT is a function defining water temperature, and INCREASED is a function defining the amount of cold water.

Four constituents are required for fuzzy control systems: providing transformation from defined physical process data to its fuzzy domain equivalent, employing linguistic statements, instituting inference for decisions, and connection to the physical process. Fuzzification interface functions map input variable values from process measurements and estimates into preset fuzzy sets, employing membership functions in the interval 0 to 1, to constitute a knowledge base. The knowledge base operator interpretation of the process to be controlled consists of the principal components of linguistic membership functions. This linguistic structure thus contains rules characterizing the goals and control strategies for the process. Logic functions operate within the linguistic structure to emulate decision making and initiate control actions through defuzzification, whose algorithms calculate the output center of gravity for actuations such as center of range and center of maximum. Rule-based fuzzy controllers thus provide a direct solution for the control of nonlinear systems as they are functionally nonlinear and time-invariant feedback controllers. They may also be employed as linear controllers that in some applications surpass the performance of PID controllers. Expert operator knowledge encoded in linguistic statements permit fast execution times because operation is not mathematically intensive. In practice, fuzzy feedback controllers providing closed-loop bandwidths in the megahertz range are routine. In contrast, PID controller closed-loop bandwidths typically are constrained to the subhertz range, owing to their integral term I required to minimize controlled variable steady-state error.

Tribological lubricant films for spacecraft applications require vacuum qualified structural bonding to sliding surfaces. These include steerable antennas, photovoltaic arrays, and gimbaled thrusters whose lubricants are manufactured employing laser ablated condensed matter. Molybdenum disulfide deposition provides an effective crystalline planar surface of weak Van der Waals forces enabling 0.01 microsliding from −200 to +700°F in vacuum. This molecular structure is shown by Figure 8-19. A laser deposition process capable of meeting

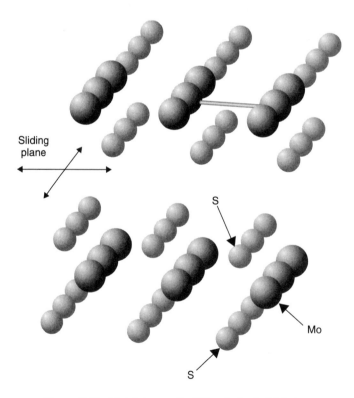

Figure 8-19. Molybdenum disulfide tribological lubricant.

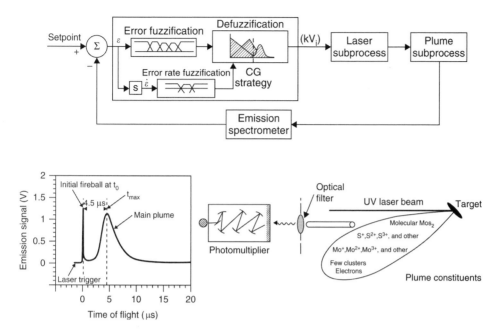

Figure 8-20. Fuzzy logic spectrometer laser control.

PF	Positive full	
PH	Positive half	
Z0	Zero	
NH	Negative half	
NF	Negative full	

		Output				
Error	P	PF	PH	NH	NH	NF
rate	Z	PF	PH	Z0	NH	NF
	N	PF	PH	PH	NH	NF
		NL	NS	ZE	PS	PL
		Error				

P	Positive
Z	Zero
N	Negative

PL	Positive large
PS	Positive small
ZE	Zero
NS	Negative small
NL	Negative large

Figure 8-21. Fuzzy logic knowledge base.

product goals uses 248 nanometer UV excimer laser energy of 300 mJ focused on a 15 mm^2 chemical target footprint in an apparatus chamber at a vacuum of 10^{-8} Torr. Efficient deposition of molybdenum disulfide product is obtained by filtered emission spectral line sensing of plume chemical constituents. That provides in-situ feedback control for critical chemical constituent spectra maximization by wideband fuzzy logic manipulation of laser cavity kilovoltage. This feedback process controller is shown in Figure 8-20. The fuzzy logic controller knowledge base is described by Figure 8-21, where the stabilizing error ε and controller response widening error-rate derivative term enhance product yield.

PROBLEMS

8-1. Robust process design for manufacturability ideally requires uncoupled, or at the least decoupled, intersubprocess parameters by process synthesis/restructuring to minimize covariance terms of subprocess mapping matrices. Alphanumerically describe mapping matrix terms for the chemical vapor deposition automation influence diagram, Figure 8-22, employing Figure 8-11, and identify the intersubprocess coupling relationship for each of the three subprocesses.

8-2. A three-zone temperature process is to be decoupled for greater effectiveness by tuning successive zone PID controllers to obtain wide-to-narrow process closed-loop bandwidths. Consider a process gain $K_p = 0.1$ for each zone in Figure 8-23, controller integral terms I operative only infrequently when controller error signals are near dc, and derivative terms turned off. Determine the respective zone controller K_c gain values required to provide closed-loop bandwidths that are separated by a factor of 10, whereby no K_c choices are permitted that are less than unity gain.

8-3. The utilization of ceramic matrix composites includes accommodating aircraft engine internal temperatures to 2500°F employing methyltrichlorosilane (MTS) chemical vapor infiltration (CVI) processing of silicon carbide preforms. Design and diagram input and output parameters for essential hierarchical subprocess controller functions using the architecture of Figure 8-10.

   Consider two apparatus subprocess feedback regulation loops for microwave preform heating and MTS infiltration gas in response, respectively, to sensed microwave power and MTS flow actuation in Figure 8-24. Then define in-situ subprocess controller X-ray measurement of preform density, plus optical-fiber-sensed infrared preform temperature, that separately out-

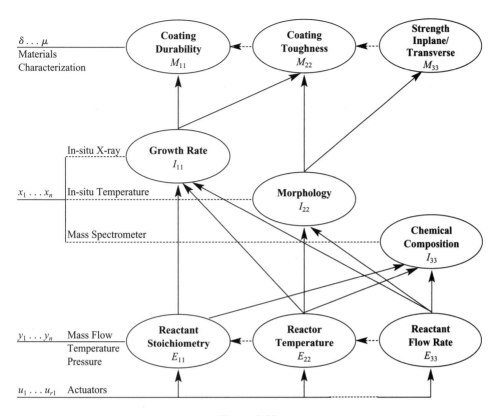

Figure 8-22.

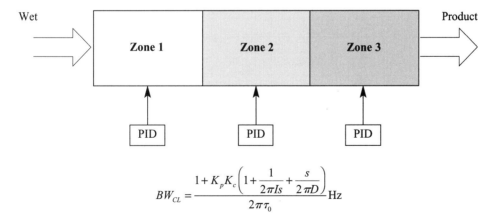

$$BW_{CL} = \frac{1 + K_p K_c \left(1 + \dfrac{1}{2\pi I s} + \dfrac{s}{2\pi D}\right)}{2\pi \tau_0} \, \text{Hz}$$

Figure 8-23.

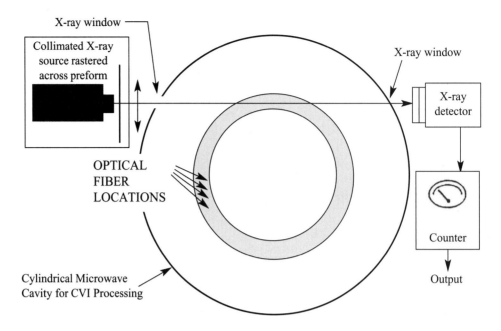

Figure 8-24.

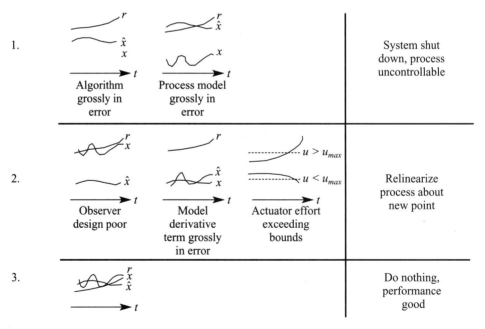

Figure 8-25.

put apparatus controller references v_1 and v_2, respectively, for MTS gas flow and preform microwave heating. Also illustrate ex-situ planner subprocess reference outputs r_1 and r_2 for preform density and temperature.

8-4. Process operations often require information not contained in process measurements, such as performance constraints or recovery procedures following process limit cycling. A process understanding model accordingly may be represented by a symbol structure that mimics processing situations employing expert rules. Construct IF, THEN, ELSE rules that describe deposition subprocess operation based upon the processing situations shown in Figure 8-25.

BIBLIOGRAPHY

1. Biggers, R.R. et al., "Emission Spectral-Component Monitoring and Fuzzy Logic Control of Pulsed Laser Deposition," *Engr. Applic. of AI* (11), 1998, 627–635.
2. Bobrow, D.G., *Qualitative Reasoning About Physical Systems,* MIT Press, 1985.
3. Boumis, P.V., *Chromate Application Improvement Options for Steel Galvanizing,* M.S. Thesis, Elec. and Comp. Engr., Univ. of Cincinnati, 1996.
4. Garrett, P.H. et al., Qualitative Process Automation For Aerospace Composites, U.S. Patent 5,032,525. 1991.
5. Garrett, P.H., Limestone Scrubber Slurry Automatic Control Systems, U.S. EPA Office Research & Development, Final Report R 805758, Research Triangle Park, NC, 1980.
6. Griffin, M.F., *Method of Obtaining the Transfer Relationship Between Scrubber Stoichiometric Ratio and pH,* BSEET Sr. Thesis, Univ. of Cincinnati, 1978.
7. Heyob, J., *The Process Discovery Autotuner,* M.S. Thesis, Elec. and Comp. Engr., Univ. of Cincinnati, 1991.
8. Jones, J.G., Intelligent Process Control of Fiber Chemical Vapor Deposition, Ph.D. Dissertation, Elec. and Comp. Engr., Univ. of Cincinnati, 1997.
9. Jones, J.G., Jero, P.D., and Garrett, P.H., "Insitu Control of Chemical Vapor Deposition for Fiber Coating," *Engr. Applic. of AI* (11), 1998, 619–626.
10. Kokar, N.M., and Reveliotis, S.A., "Integrating Qualitative and Quantitative Methods in Model Validation and Monitoring," in *IEEE Symp. Intell. Control,* Philadelphia, 1991.
11. LeClair, S.R., Abrams, F.L., and Matejka, R.F., "Qualitative Process Automation: Self-Directed Manufacture of Composite Materials," *AI EDAM* 3(2), 1989, 125–136.
12. Moore, D.C., *Subprocess Control Design Methods for Advanced Materials Processing,* M.S. Thesis, Elec. and Comp. Engr., Univ. of Cincinnati, 1994.
13. Saridis, G.N. and Valavanis, K.P., "Analytical Design of Intelligent Machines," *Automatica,* 24(2), 1988, 123–130.
14. Suh, N.P., *Axiomatic Design,* Oxford University Press, New York, 2001.
15. Ziegler, J., and Nichols, W., "Optimum Settings for Automatic Controllers," *ASME Transactions,* November, 1942.

CHAPTER 9

Process Automation Applications

9-0 INTRODUCTION

This chapter introduces diverse applications of the methods developed in Chapter 8. Titanium forging is described in the first application utilizing a priori Ashby map behavior models of microstructure for forging temperatures and strain rates within a specified range, whereby product volume changes remain derogation free throughout processing. Measured process features for decision and control purposes are then applied to a nanomaterial product mixate whose molecular dynamics and morphology are interpreted by in-situ impedance spectrometry and the Havriliak–Negami phenomenological relationship. Archived impedance data correlated with preferred material properties are employed to achieve exfoliated morphologies employing in-situ electric field actuation to lock in the goal product as a thermoset nanocomposite.

Laser ablated deposition that achieves stoichiometric condensed matter enables the production of engineered crystalline products, including superconductors. Empirical sensor data of mass-momentum-energy process quantities provide online system identification observations for ex-situ planner modeling. The resulting feedforward model directs this ablation process over a goal product trajectory with uniform deposition until a rule-based parameter-consensus detector of performance declination initiates system identification remodeling.

Steel recrystallization annealing is demonstrated next, employing an ex-situ planner ANNEAL NET that has been trained offline from exemplar punched-

Advanced Instrumentation and Computer I/O Design, Second Edition. By Patrick H. Garrett
Copyright © 2013 the Institute of Electrical and Electronics Engineers, Inc.

steel-strip coupons with parameters that include Rockwell hardness. This network thus learns from experience to direct a gradient-descent algorithm identifying ten annealing zone temperature values for attenuating strip hardness variance. A final application illustrates compensation of molecular beam epitaxy machine nonlinear interactions, through ultralinear flux calibration, to enhance ellipsometry-sensed process features for semiconductor growth control.

9-1   ASHBY MAP GUIDED EQUIAXED TITANIUM FORGING

Titanium alloys are required for critical aerospace structures and propulsion components that rely upon accurate thermomechanical processing for forged strength. Progress in utilizing material behavior models to optimize Ti–6Al–4V alloy properties, including the interrelation of microstructure with forging temperature and strain rate, is demonstrated. The described processing state sequences permit defect-free component manufacture with hot working absent strain-induced porosity. Forged component quality maturity results from the breakup of lamellar microstructure into an equiaxed grain structure through high-speed deformation. This is directed by a material microstructure feedforward planner Ashby map providing quantitative forging-state direction.

Microstructure features that alleviate product derogation during forming is a feature of this forging process. The thermomechanical deformation Ashby map shown by Figure 9-1 provides a knowledge base for ex-situ planner feedforward control. Defect-free equiaxed microstructure employing high-speed deformation was discovered from dislocation-induced nucleation of a lower temperature alpha microstructure phase [24]. Titanium alloys are difficult to work because of microstructural complexity, but strain energy associated with volume change is beneficial to equiaxed alpha precipitates.

Process control is directly actuated during forging-state execution to achieve titanium property realization by providing the required apparatus controller functions, principally temperature regulation and ram velocity motion control, that maintain deformation strain rates above 1/second. This is illustrated by the subprocess influence diagram of Figure 9-2, where forging actuation is absent in-situ microstructure feedback control because of the unavailability of practical sensors. In substitution, an in-situ compensator achieves goal strain and temperature profiles by providing apparatus controller references for feedback regulation of their respective sensed forge apparatus parameters, including forge response.

Ti–6Al–4V represents more than 50% of total titanium alloy components worldwide. Its material properties during manufacture are sensitive to process in-

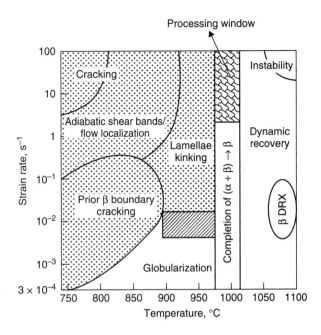

Figure 9-1. Ti-6Al-4V microstructure Ashby map.

fluences, including forging temperature, strain rate, and processing-state history, but microstructure evolution is sufficiently well defined that planner maps can achieve accurate forging execution. Further, aluminum addition increases alloy strength at lower temperature forging whereas vanadium increases ductility at higher forging temperatures. Apparatus parameter upper limits include 1100°C and strain rates to 10/second. Kinetic breakup of lamellar microstructure, including precipitation into equiaxed microstructure, improves reliability by reducing crack initiation with product fatigue loading.

9-2 Z-FIT MODELED SPECTRAL CONTROL OF EXFOLIATED NANOCOMPOSITES

Engineered composition and morphology properties that can be tuned are especially of value for meeting nanomaterials processing requirements. The manufacturing system described here illustrates processing of inorganic/organic nanomaterials such as high-strength, light-weight artificial bone prosthesis. Nanomaterials property realization requires process apparatus with sufficient measurement accu-

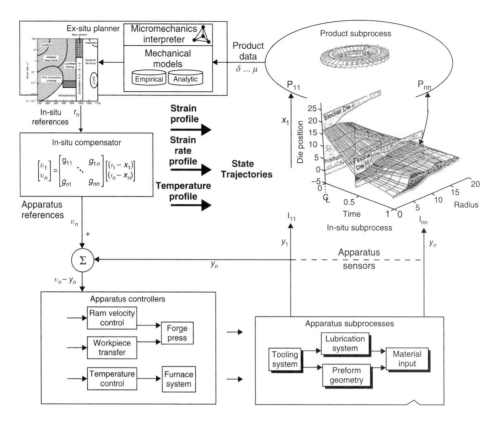

Figure 9-2. Equiaxed titanium forging subprocess influences.

racy and control capability to meet product composition and structure goals. The system described provides advancement in dielectric impedance sensing to permit access to material states and properties previously unresolvable. Morphology may be sufficiently interpreted by dielectric impedance spectrometry parameter measurements to provide characterization of evolving states within material interfacial regions and enable in-situ process automation. Material mixtures of interest include EPON 862 polymer matrix nanofiller host, 1–6% by weight Nanocore 130E nanoclay montmorillonite mineral, and Epikure W curing agent (shown in Figure 9-3).

A Novocontrol spectrometer whose bandwidth extends from 30 μHz to 20 MHz can analyze multiple electrical parameters resolvable from material dielectric charge and molecular dynamics. These correlate with product morphology features ranging between exfoliated (good) and agglomerated (bad). Materials molecular and ionic behavior typically span broad frequency and impedance

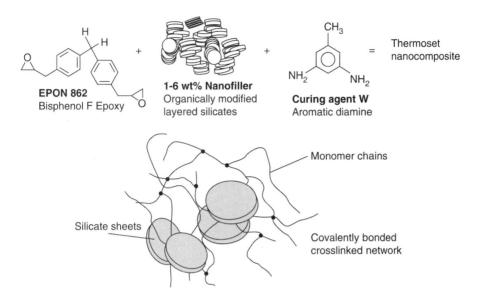

Figure 9-3. Clay epoxy nanocomposite material properties.

ranges, placing a significant performance demand on dielectric spectrometers. Nanoclays also uniquely exhibit strong polarization accompanied by dielectric relaxation. For example, at the start of the cure process, the low-viscosity mixture hardens, with ensuing crosslinked reactions measurable as progressively diminishing ionic activity at the end of the curing. Figure 9-4 shows a snapshot of the spectrometer α-relaxation parameter shape, resulting from interfacial polarization dynamics spanning from below 1 Hz to 1 MHz, that identifies goal exfoliated morphology for a material glass temperature of 200°C.

The Havriliak–Negami relationship offers a phenomenonological model supporting dielectric impedance interpretations whose utility is aided by the synthesis of complex nonlinear least squares (CNLS) data fitting. In application, Z-Fit software iterates archived product impedance data to identify real-time spectrometer-sensed features equivalent to material properties of interest [11]. This model-fitting procedure also employs the Levenberg–Marquardt nonlinear least squares technique for locating the minima of multivariate functions.

Nanocomposite processing described by Figure 9-5 describes both quantitative data furnished by Z-Fit feedforward planner models and empirical data provided by online impedance spectrometer observations. That enables externally applied electric field directed actuation to physically lock in a matching goal exfoliated nanomaterial morphology of interest, which is preserved by autoclave

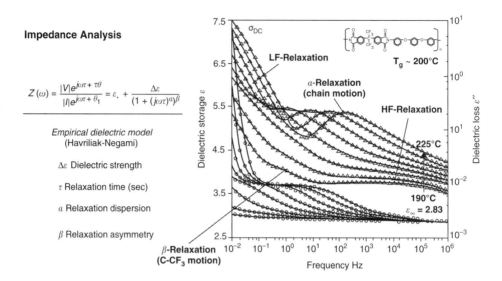

Figure 9-4. Nanocomposite spectrometry Z-fit model.

cure physiochemical reactions by converting it to a nanocomposite, with beneficial cured product decoupling from agglomeration morphology influences. The benefits of decoupled subprocess influence control are retained, including feedforward planner attenuation of long-time-constant disturbances, impedance spectrometer in-situ control of short-time-constant disorder, and process stabilization through apparatus parameter regulation. Corroborating the physical morphologies of manufactured products requires laboratory evaluation, including transmission electron microscopy (TEM) and azmuthal X-ray small angle (AXSA) diffraction.

9-3 SUPERCONDUCTOR PRODUCTION WITH ADAPTIVE DECISION AND CONTROL

Yttrium barium copper oxide (YBCO) is a ceramic superconductor material with a unit cell structure shown by Figure 9-6 and a critical temperature of 92°K that is ideally suited for lossless spaceborne power systems. However, laser ablation processing of the ideal compound $YBa_2Cu_3O_{6.85}$ is sensitive to stoichiometric variations requiring the versatility of in-situ process control methods for consistent manufacture. A practical model of laser ablation deposition that characterizes energy transformations is shown in Figure 9-7; its utility arises from a shortage of defining laser ablation process models. Coherent ultraviolet laser fluence creates

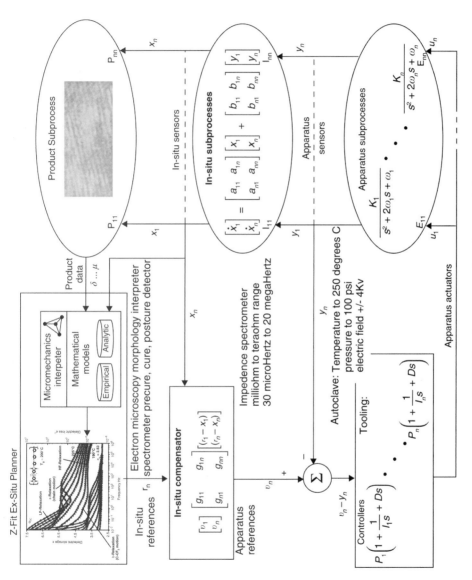

Figure 9-5. Exfoliated nanocomposite subprocess influences.

251

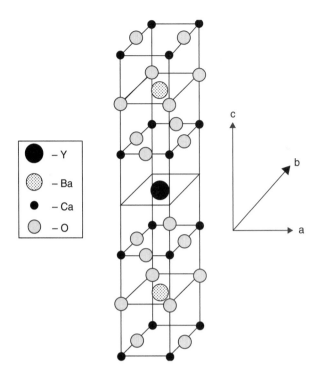

Figure 9-6. YBCO orthorhombic superconductor.

energetic electric and kinetic molecular plume species, including intermediate chemical reactions whose substrate deposition may also include recombinant material. Plume density is modeled by the particle distribution equation shown in Cartesian basis functions of material distribution [23].

Creating an effective ex-situ planner for continuous superconductor production is difficult also because of the scarcity of laser ablation models. This identification challenge is met by a process data observer through empirical measurements of separate microbalance deposition growth, spectrometer plume species, and incident laser energy sensors. Acquired mass, momentum, and energy system identification data are then imported into an adaptive observer in-situ state estimator diagrammed by the hierarchical subprocess influences of Figure 9-8. Processing complexity is reduced by the natural subprocess decoupled parameter occurrence whose matrix definitions are tabulated in Table 9-1.

Laser deposited material sensing of process substrate thickness and thickness rate permit deposited film microstructure acquisition, using a quartz crystal microbalance positioned within the plume plasma without shadowing the substrate.

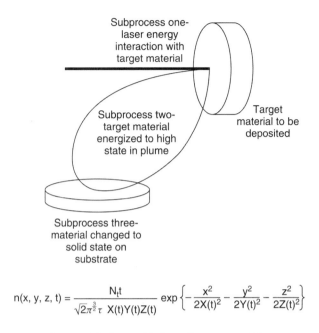

$$n(x, y, z, t) = \frac{N_t t}{\sqrt{2}\pi^{\frac{3}{2}}\tau \; X(t)Y(t)Z(t)} \; \exp\left\{-\frac{x^2}{2X(t)^2} - \frac{y^2}{2Y(t)^2} - \frac{z^2}{2Z(t)^2}\right\}$$

Figure 9-7. Laser ablation subprocess model.

An Inficon XTC sensor indicates thickness t_f between 10 and 10,000 Å by crystal frequency changes resulting from deposited mass buildup according to the algorithm of Equation (9-1). Offline scanning electron microscopy characterization of deposited films verified a typical thickness variability of approximately 3%FS, providing a resolution 2^{-n} of 0.03 corresponding to five-bit binary accuracy. Second, optical emission spectroscopy of the plume permits chemical line-spectra measurement with wideband digitization at 400 megasamples per second (Table 4-2) 1.5 μs width, −3dB plume amplitudes. Data acquired in terms of step-interpolated signals are evaluated for its 2^{-n} amplitude fraction, referenced to unity full scale, corresponding to seven-bit binary accuracy by Equation (9-2). Third, a laser bolometer power sensor possesses a mean linearity of $\overline{0.1}$%FS whose signal is digitized for input to the laser controller, having a dominant 1σ random error of 0.11%FS. The combined mean and random measurement error of 0.21%FS is evaluated as the 2^{-n} value of 0.0021 equivalent to nine-bit binary accuracy by Equation (9-3). These acquired system identification data, acconpanied by their respective accuracies, are described in Figure 9-9. Evaluation of binary accuracy for each quantity is aided by their decimal equivalent identities from Table 5-7 of Chapter 5. The sensor suite for this superconductor process is illustrated by Figure 9-10.

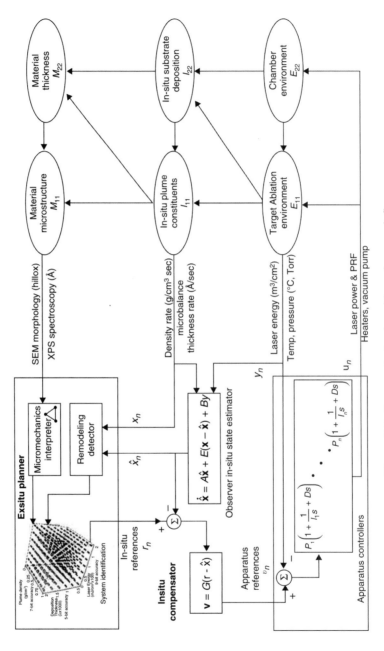

Figure 9-8. Superconductor deposition subprocess influences.

Table 9-1. Decoupled superconductor subprocess parameters

$$\begin{bmatrix} \text{SEM Morphology (hillox)} \\ \text{XPS Spectroscopy (Å)} \end{bmatrix} = \begin{bmatrix} M_{11} & 0 \\ M_{21} & M_{22} \end{bmatrix} \begin{bmatrix} \text{Density rate (g / cm}^3 \text{ sec)} \\ \text{XPS Spectroscopy (Å/sec)} \end{bmatrix}$$

$$\begin{bmatrix} \text{Density rate (g / cm}^3 \text{ sec)} \\ \text{Microbalance thickness rate (Å / sec)} \end{bmatrix} = \begin{bmatrix} I_{11} & 0 \\ I_{21} & I_{22} \end{bmatrix} \begin{bmatrix} \text{Laser energy (mJ / cm}^2) \\ \text{Temp, pressure (°C, Torr)} \end{bmatrix}$$

$$\begin{bmatrix} \text{Laser energy (mJ / cm}^2) \\ \text{Temp, pressure (°C, Torr)} \end{bmatrix} = \begin{bmatrix} E_{11} & 0 \\ 0 & E_{22} \end{bmatrix} \begin{bmatrix} \text{Laser power and PRF} \\ \text{Heaters, vacuum pump} \end{bmatrix}$$

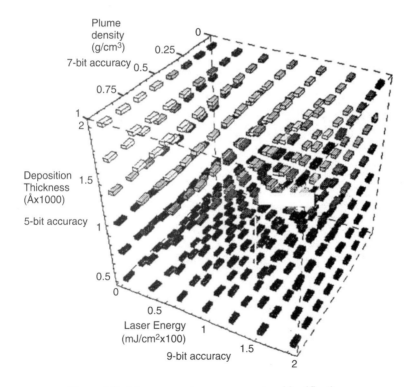

Figure 9-9. Mass-momentum-energy system identification.

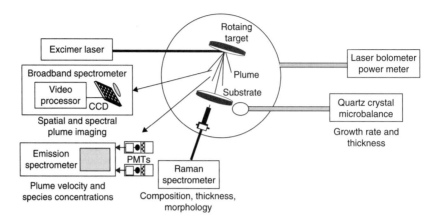

Figure 9-10. Superconductor process sensor suite.

$$t_f = \left[\frac{N_q d_q}{\pi d_f f_c C} \right] \text{ thickness algorithm} \qquad (9\text{-}1)$$

t_f = film thickness
d_q = quartz density (g/cm³)
N_q = crystal frequency constant (Hz/cm)
d_f = film density (g/cm³)
f_c = coated crystal frequency (Hz)
C = calibration constant (1/cm²)

$$2^{-n} = 0.03 \text{ thickness resolution}$$

$$n = |3.32 \log (2^{-n})|$$

$$= \text{five bits ex-situ thickness accuracy}$$

$$2^{-n} = \frac{\sqrt{2} \; \pi \, \text{plume BW}}{\sqrt{5} \, \text{sampling } f_s} \qquad (9\text{-}2)$$

$$= \frac{\sqrt{2} \; \pi \, (2 / 1.5 \, \mu\text{sec width})}{\sqrt{5} \; 400 \text{ megasamples}}$$

$$= 0.0066 \text{ plume amplitude resolution}$$

$$n = |3.32 \log (2^{-n})|$$

$$= \text{seven bits in-situ plume accuracy}$$

$$2^{-n} = 0.0021 \text{ laser energy resolution} \qquad (9\text{-}3)$$

$$n = |3.32 \log (2^{-n})|$$

$$= \text{nine bits laser energy accuracy}$$

Ex-situ planner structures incorporate feedforward production models to provide resolved event-based in-situ control references along process state trajectories. Planners are essential because process apparatus regulation is generally capable of only limited achievement of product goals. Although ex-situ planner product representations are highly definitive, they are difficult to realize for laser deposition processing because of variability in product property translation and laser ablation efficiency. Process nonlinearity is least problematical for laser target apparatus subprocesses, which provide outputs y_n in response to actuation inputs u_n. The in-situ plume subprocess is more nonlinear and critical to deposition performance. It consequently requires a more comprehensive control implementation to achieve product specifications.

The superconductor adaptive observer-based in-situ planner employs measured mass, momentum, and energy system identification data to execute in-situ feedback control, as diagrammed in Figure 9-8. In-situ subprocess parameters x_n are controlled in response to ex-situ planner executed in-situ references r_n and the in-situ compensator G matrix, as influenced by the observer feedback controller E matrix and apparatus measurements y_n. The merits of the observer in-situ state estimator also include attenuation of process noise and disturbance propagation for improved in-situ control performance, where noiseless estimated in-situ parameters $\hat{x}_n$ are employed in the control algoriithm instead of measured values x_n. Deviation between these values constitutes migration of actual from ideal superconductor production, whose detection initiates remodeling of the E matrix to achieve reconvergence of $\hat{x}_n$ with x_n. Remodeling initiation is described by the consensus rule base of Figure 9-11 [14].

Ex-situ planner identification remodeling is updated employing radial basis function Equations (9-4) and (9-5), respectively, defining thickness growth rate m and plume density a for redirecing setpoint control of laser energy e and repetition rate p. These parameters are imported into remodeling linear partial-differential Equation (9-6) to reacquire in-situ thickness and plume density states that restore superconductor production rates. Repeated remodeling is exercised with continued production migration. It is notable from the system identification example of Figure 9-9 that maximizing plume density indices at 0.75 and deposition thickness at 2.0 occur at a laser energy index of 0.5, corresponding to one-fourth of the energy available.

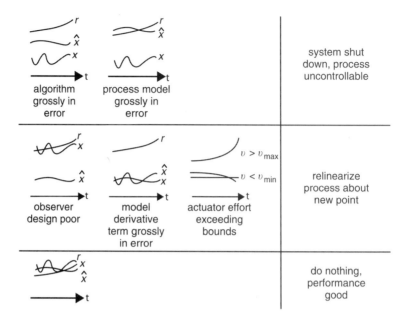

Figure 9-11. Remodeling consensus rule base.

Thickness:

$$\dot{m} = \Sigma k_m \exp\left\{ -\frac{|m - \mu_{mh}|^2}{2\sigma_{mh}^2} - \frac{|a - \mu_{ah}|^2}{2\sigma_{ah}^2} - \frac{|e - \mu_{eh}|^2}{2\sigma_{eh}^2} - \frac{|p - \mu_{ph}|^2}{2\sigma_{ph}^2} \right\} \quad (9\text{-}4)$$

Plume:

$$\dot{a} = \Sigma k_a \exp\left\{ -\frac{|m - \mu_{mh}|^2}{2\sigma_{mh}^2} - \frac{|a - \mu_{ah}|^2}{2\sigma_{ah}^2} - \frac{|e - \mu_{eh}|^2}{2\sigma_{eh}^2} - \frac{|p - \mu_{ph}|^2}{2\sigma_{ph}^2} \right\} \quad (9\text{-}5)$$

Remodeling:

$$\begin{bmatrix} \dot{m} \\ \dot{a} \end{bmatrix} = \begin{bmatrix} f_1(m,a,e,p) \\ f_2(m,a,e,p) \end{bmatrix} = \left.\frac{\partial f}{\partial(m,a)}\right|_{\substack{m_0,a_0 \\ e_0,p_0}} \begin{bmatrix} \Delta m \\ \Delta a \end{bmatrix} + \left.\frac{\partial f}{\partial(e,p)}\right|_{\substack{m_0,a_0 \\ e_0,p_0}} \begin{bmatrix} \Delta e \\ \Delta p \end{bmatrix} \quad (9\text{-}6)$$

where
m = microbalance sensed thickness (Å)
a = spectrometer sensed plume density (g/cc)
e = laser energy density (mJ/cm^2)
p = laser pulse repetition rate (Hz)

9-4   NEURAL NETWORK ATTENUATED STEEL ANNEALING HARDNESS VARIANCE

Steel recrystallization annealing is demonstrated employing neural-network-directed ductility reconstitution. This is achieved by means of ex-situ planner ANNEAL NET that has been trained offline with strip coupon parameters, including Rockwell hardness. The trained network provides direction of a gradient descent algorithm identifying ten annealing-zone temperature values for attenuating strip hardness variance. The annealing process accommodates cold-reduced steel strip in motion with thermal modification of its crystalline grain microstructure to within two Rockwell units of 60 [22]. Wheeling-Pittsburgh Steel Corporation provided their annealing works, shown by Figure 9-12, including 3519 separate strip coupon parameter samples for network training purposes.

Thermal annealing for restoration of cold-reduced steel strip ductility is a common steel process operation amenable to improvement by means of computationally intelligent automation. Limitations from the significant size of the physical apparatus restrict process modifications, but opportunities are available with the multiple adjustable process parameters. Improving product value by discovering process compensation beyond that available from trial and error is an objective. Outcomes of this effort include a neural-network-based ex-situ planner incorporating definitive product property modeling, and in-situ strip temperatures are coincident with apparatus zone temperatures, simplifying zone heating control owing to the favorable ratio of strip surface area to its volume. This is illustrated by the subprocess influences of Figure 9-13.

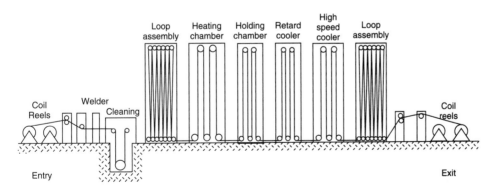

Figure 9-12. Steel recrystallization annealing apparatus.

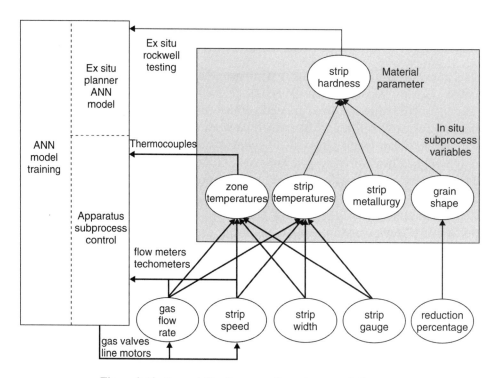

Figure 9-13. Recrystallization annealing subprocess influences.

Steel recrystallization annealing improvement benefits from product property modeling employing defining steel strip ductility from testing similar previous samples. Cold-reduced steel strip exhibits grain elongation accompanied by increased strength and hardness, but a decrease in ductility that limits its value for product forming such as stamping operations. Fortunately, reheating cold rolled steel to an example recrystallization temperature of 750°F restores ductile microstructure properties. However, for continuous moving strip minimizing end-to-end hardness variance by varying specific heat application to the strip is challenging, often relying on trial and error. Rockwell hardness evaluation is an ex-situ mechanical measurement of strip deformation resistance obtained from test coupons punched from exemplar product. A Rockwell differential indentation measurement is beneficial to evaluation accuracy. Steel strip annealing has a typical processing goal of Rockwell hardness of 60.

Production processes require appropriate models to direct controlled variables along ideal proces state trajectories in order to achieve product goals. This assumes the existence of capable process and control apparatus as well as robust product property modeling. Two considerations are invoked for simplification of the recrystallization annealing ex-situ planner: time-invariant process behavior and strip hardness independence from metallurgical composition. A neural network proces model designated ANNEAL NET was structured to provide one output parameter of Rockwell hardness for thirteen process parameter inputs, including strip gauge, strip width, strip speed, and ten apparatus zone temperatures.

ANNEAL NET is a feedforward network with one hidden layer employing back propagation for training weight values w that incorporate a momentum factor α to accelerate learning. The structure used for this design is shown in Figure 9-14. Back propagation is a learning procedure that calculates the derivatives of the output for a nonlinear differentiable system, with respect to all inputs and parameters of the system, through calculations that proceed backward from outputs to inputs. A bipolar sigmoidal activation function $f(S)$ ensures the ability of the network to accommodate the large number of inputs presented, where its derivative is implemented as $f'(S) = \lambda/2[1 - f(S)^2]$, with λ regulating steepness. The hidden layer consists of five neurons. Training error is shown employing 3519 product coupons in Figure 9-15, each consisting of measured hardness, with a deviation error of two Rockwell hardness units predicted. The use of excessive training coupons provides network overtraining with no additional benefit.

$$f(S) = \frac{1}{1 - e^{-\lambda s}} - 1 \qquad \text{bipolar sigmoid} \qquad (9\text{-}7)$$

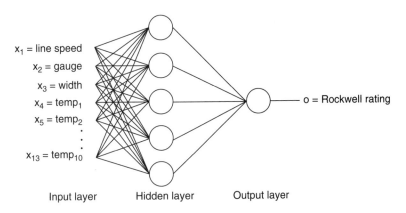

Figure 9-14. ANNEAL NET model architecture.

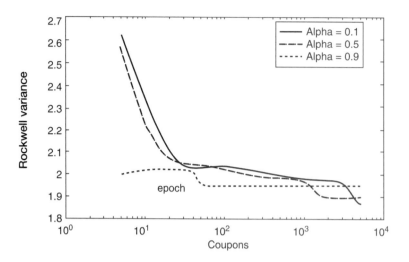

Figure 9-15. ANNEAL NET Rockwell variance versus coupon training.

Ex-situ planner zone temperature determination realizes operating temperature values based on production strip guage, width, and line-speed inputs. These parameters provide convergence to a goal hardness h by network output hardness o, using updated estimates employing a gradient descent algorithm. The ten zone temperatures estimated provide apparatus temperature setpoints. A relative cost function error C guides this evaluation by encouraging network hardness output changes that decrease C for input data samples and zone temperature iterations. Cost function minimization is achieved through adaptation of network weights. Employing a sensitivity derivative enables the gradient of the network output to be fed back as estimates for back propagation training. Compromise between learning momentum and error minimization is realized by the weight iteration equation plotted for three alpha values in Figure 9-15. The gradient descent algorithm is shown by Figure 9-16.

$$C = \frac{1}{2}(o - h)^2 \qquad \text{cost function} \qquad (9\text{-}8)$$

$$\frac{\partial C}{\partial x} = -(h - o)\frac{\partial o}{\partial x} \qquad \text{sensitivity derivative} \qquad (9\text{-}9)$$

$$\Delta w(i) = \frac{\partial C}{\partial w(i)} + \alpha \Delta w(i-1) \qquad \text{weight iteration} \qquad (9\text{-}10)$$

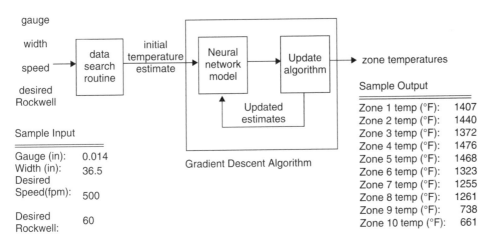

Figure 9-16. Gradient descent neural directed zone temperatures.

9-5 ULTRALINEAR MOLECULAR BEAM EPITAXY FLUX CALIBRATION

The molecular beam epitaxy machine was developed by Al Cho at Bell laboratories during the 1960s for the production of multilayer optical and electronic devices. MBE offers versatile growth of semiconductor material whose value continues for prototyping new devices, where benefits include a growth chamber at liquid nitrogen temperature and pressure of 10^{-10} Torr providing high isolation to contamination. Of interest is achieving automatic control of the MBE laboratory process with the robustness of industrial process control. It is consequently of interest to seek MBE machine enhancement by decoupling cell flux effusion from sources of apparatus variability, disturbance singularities, and processing disorder to aid ellipsometry sensed features for growth control. Figure 9-17 illustrates typical process and control apparatus for an MBE machine.

Molecular beam epitaxy semiconductor processing has been advanced through ultralinear flux calibration described by Figure 9-18. This aids electrooptical devices grown in multilayer thin-film forms and production performance overall. Nominal MBE variability may be achieved by substituting trapezoidal PID cell temperature controller tuning for classical tuning methods. A second improvement is feedforward planner compensation for flux enthalpy transient disturbances at cell openings. Third, replacement dc power drivers for conventional cell triac ac power drivers to eliminate flux disorder arising from electric utility power frequency corrections. These improvements beneficially enhance spectro-

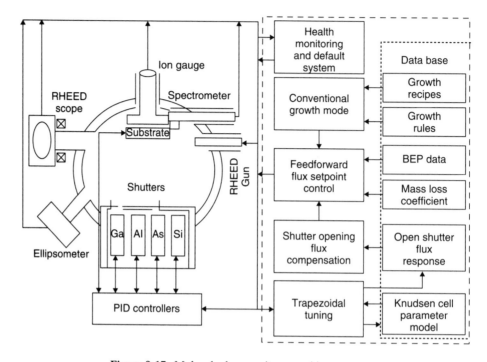

Figure 9-17. Molecular beam epitaxy machine apparatus.

scopic ellipsometry in-situ control measurements of semiconductor growth properties, including achievement of multicell flux synchrony within 1%FS at 1 σ variance.

MBE machines employ tracking of cell temperature values correlated with flux F by Equation (9-11), while compensating for cell material loss as consumed, to define a beam equivalent pressure (BEP) of F. Cell temperature accuracy and controller tuning are therefore critical to this flux inference, since during material growth the cell flux ion gauge must be repositioned to permit unobstructed growth. Experimental observation has shown that flux accuracy maintenance requires retuning cell temperature controller PID values for temperature setpoint changes exceeding 50°C. Controller tuning must achieve unconditional stability, minimum steady-state error between cell and setpoint temperature values, and unvarying flux following cell load changes. These conditions are determined by correct controller gain P, integral time I, and derivative time D determination. Figure 8-4 demonstrates a half-order-of-magnitude reduction in temperature steady-state error for trapezoidal compared with quarter-decay controller tuning [8].

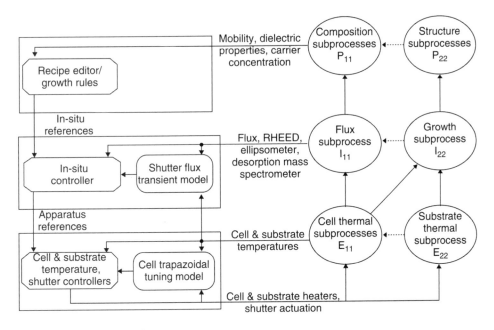

Figure 9-18. MBE ultralinear calibration subprocess influences.

$$F = \frac{A \cos \theta\ P}{\pi r^2 \sqrt{2\pi m k T}} \tag{9-11}$$

where
F = Flux atoms/cm^2 at wafer
A = Knudsen cell aperture area
θ = Cell–wafer offset angle
P = Flux pressure at wafer
T = Knudsen cell temperature
r = Shortest cell–wafer distance
k = Boltzman's constant
m = Mass of effluent

Cell element evaporants each have a finite thermal mass such that their respective fluxes respond to both heater excitation temperature input changes with time constant τ_0, typically one second, and an evaporant output flux time constant at cell opening of τ_2, observed as 20 seconds in Figure 9-19. Figure 9-20 proceeds to describe a feedforward temperature compensation algorithm that provides a

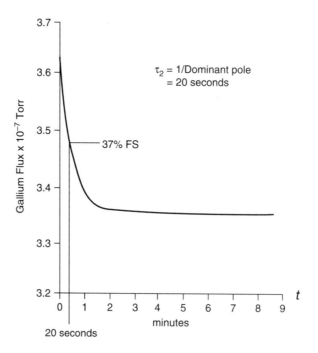

Figure 9-19. Cell opening flux time constant.

cell temperature setpoint increase, by Equation (9-12), prior to cell opening to cancel the flux transient droop. The initiation time for this compensation prior to cell opening is defined by the trapezoidal process lag measurement of Figure 8-4. Flux disturbance correction of beam flux values from 10% to 1% may be achieved with this compensator.

$$+\Delta R_k = \frac{+\Delta R_k' \cdot \Delta\text{Flux}}{0.9\,\Delta\text{Flux}'} \cdot \left[1 + \left(\frac{\tau_2' - \tau_2}{\tau_2}\right)\right] \cdot \exp(-t/\tau_2) \qquad (9\text{-}12)$$

where

$+\Delta R_k$ = Temperature setpoint increment for shutter transient compensation initiated at t_0 minus process lag time.

ΔFlux = Shutter opening flux change to steady state.

τ_2 = Evaporant mass time constant from ΔFlux.

$+\Delta R_k'$ = Arbitrary temperature setpoint increment for open shutter bump test.

$\Delta\text{Flux}'$ = Open shutter flux change to steady state for $+\Delta R_k'$.

τ_2' = Evaporant mass time constant from $\Delta\text{Flux}'$.

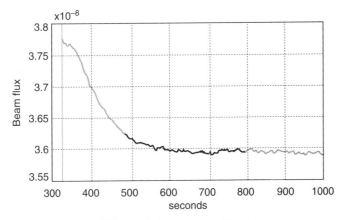

(a) Flux at shutter opening without shutter
opening transient compensation

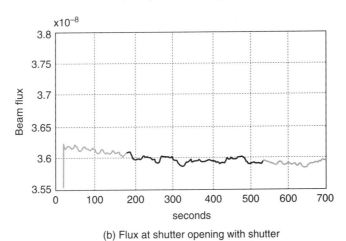

(b) Flux at shutter opening with shutter
opening transient compensation

Figure 9-20. Cell enthalpy flux transient feedforward compensator.

Online process data further reveals flux amplitude disorder up to 8% of average flux values at 10-minute intervals. Analysis discloses that electric power utilities execute millihertz frequency corrections every 10 minutes to compensate for generator speed variations for users that rely on counting power cycles for timekeeping. MBE machine flux disorder arises from these corrections by two circumstances. One is the common use of triac ac power drivers for cell temperature heaters. The second is a typical flux evaporant cell bandwidth of 0.008 Hz, evaluated by Equation (9-13) that permits transmission of millihertz frequency power corrections to flux outputs. Decoupling of power frequency corrections is obtained by substitution of programmable 1-kW dc power drivers for common triac ac drivers. In addition to decoupling power frequency corrections from cell flux sources, the instantaneously available power density of direct current provides improved control response.

$$BW = \frac{\text{Dominant pole}}{2\pi} \quad \text{Cell bandwidth} \qquad (9\text{-}13)$$

$$= \frac{1}{2\pi\tau_2}$$

$$= \frac{1}{(2\pi)(20\,\text{sec})} \quad \text{for Ga}$$

$$= 0.008\ \text{Hz}$$

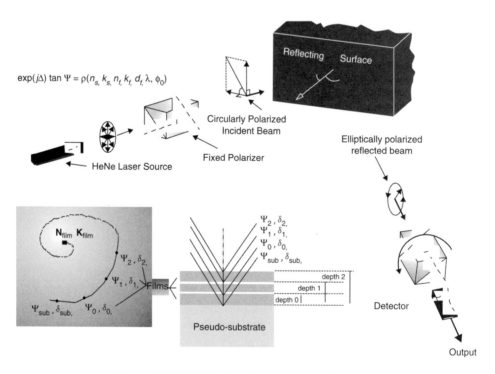

Figure 9-21. MBE in-situ ellipsometry growth sensor.

The foregoing MBE machine flux calibration assists semiconductor production accuracy by enabling multiple wavelength spectroscopic ellipsometry. Sensing relies upon the optics of stratified product media for nondestructive in-situ measurement of surface and substrate properties to assist mechanizing structure and composition control. This sensor system is illustrated in Figure 9-21. In practice, elliptically polarized laser emissions reflected from substrate growth are analyzed for electric field(ψ) and phase(δ) pairs versus wavelength to evaluate material properties. Substrate optical parameters *n, k,* and *d,* respectively, the index of refraction, extinction coefficient, and thickness, are then evaluated employing Fresnel equations [26].

BIBLIOGRAPHY

1. Busbee, J., Laube, S.J.B., and Jackson, A.G., "Sensor Principles and Methods for Measuring Physical Properties," *J. of Materials,* Vol. 48, No. 9, September 1996, 16–23.
2. Cho, A.Y., "Growth of III-V Semiconductors by Molecular Beam Epitaxy and Their Properties," *Thin Solid Films,* No. 100, 1983.
3. Frost, H.J., and Ashby, M.F., *Deformation Maps,* Pergammon Press, 1982.
4. Garrett, P.H. et al., "Decoupled Flux Control for Molecular Beam Epitaxy", *IEEE Transactions on Semiconductor Manufacturing,* Vol. 6, No. 4, November 1993, 348–356.
5. Garrett, P.H., Jones, J.G., Moore, D.C., and Malas, J.C., "Emerging Methods for the Intelligent Processing of Materials, *J. Matls. Engr. and Performance,* Vol. 2, No. 5, 1993, 727–732.
6. Gibson, J.R. et al., "Automatic Control System for the Continuous Annealing Line at USS-POSCO," *Iron and Steel Engineer,* Vol. 69, March 1992, 424–429.
7. Haykin, K., *Neural Networks,* Macmillan, New York, 1994.
8. Heyob, J.J. et al., A Hierarchical Control System for Molecular Beam Epitaxy, U.S. Patent 5, 461,559, July 1996.
9. Ishikawa, T. et al., "Application of Modern Control to Temperature Control of the MBE System," *Japan J. Applied Physics,* Vol. 29, No. 3, March 1990.
10. Jacobs, J.D. et al., "Toward Closed-Loop Process Control of Nanocomposites: Impedance Monitoring and E-Field Directed Morphology," in *SAMPE,* Boston, 2010, pp. 1–8.
11. Jacobs, J.D., *Online Impedance Spectrometry of Thermoset Nanocomposites for In-Situ Process Control,* Ph.D. Dissertation, Elec. and Comp. Engr., Univ. of Cincinnati, 2009.
12. Jones, J.G. et al., "Image Processing Plume Fluence for Superconducting Thin Film Deposition," *Engr. Applications of AI,* Vol. 13, No. 5, October 2000, 598–601.
13. King, P.J., and Mandani, E.H., "The Application of Fuzzy Control Systems to Industrial Processes," *Automatica,* Vol. 13, Pergammon Press, UK, 1977, p. 235.
14. Laube, S.J.P., *Hierarchical Control of Pulsed Laser Deposition for Manufacture,* Ph.D. Dissertation, Elec. and Comp. Engr., Univ. of Cincinnati, 1994.
15. Laube, S.J.P., Pulsed Laser Deposition Improvement by Self-Directed Control, USAF Technical Report, WL-TR-95-4079, May 1995.
16. Lubbers, D.P. *In-Situ Monitoring and Control of Pulsed Laser Deposition of Superconducting Films,* M.S. Thesis, Elec. and Comp. Engr., Univ. of Cincinnati, 1996.

17. Malas, J.C., et al., Microstructure Driven Design, USAF Technical Report. WL-TR-96-4109, May 1996.

18. Matejka, R.F., *A Programming Environment for Qualitative Process Control,* M.S. Thesis, Elec. and Comp. Engr., Univ. of Cincinnati, 1988.

19. Moody, J., Darken, C.J., "Fast Learning in Networks of Locally Tuned Processing Units," *Neural Computation,* Vol. 1, 1989, 281–294.

20. Murray, S.P., *Digital Control System Design and Demonstration for PLD Superconducting Films,* M.S. Thesis, Elec. and Comp. Engr., Univ. of Cincinnati, 1994.

21. Park, J., and Sandberg, I.W.,"Universal Approximation Using Radial-Basis-Function Networks," *Neural Computation,* Vol. 3, 1991, 246–257.

22. Reddy, R.A., *ANNEAL NET: Wheeling-Pittsburg Steel Hardness Variance Minimization,* M.S. Thesis, Elec. and Comp. Engr., Univ. of Cincinnati, 1996.

23. Singh, R.K., and Narayam, J, "Pulse Laser Evaporation for Deposition of Thin Films: Physics and Theory Model," *Physical Review,* Vol. 41, No. 13, May 1990, 8843–8859.

24. Tamirisakandala, S., Dutta, B, "Influence of Prior Deformation Rate on the Mechanism of Beta to Alpha + Beta Transformation in Ti-6Al-4V," *Scripta Materiala,* Vol. 46, 2002, 673–6787.

25. Tamirisakandala, S. et al., "Recent Advances in the Deformation Processing of Ti Alloys," *J. Matls. Engr. and Performance,* Vol. 12, No. 6, 2003, 661–669.

26. Tompkins, H.G. et al., *Spectroscopic Ellipsometry and Reflectrometry,* NewYork, Wiley, 1999.

27. Vaia, R.A., and Maguire, J.F.," Polymer Nanocomposites with Prescribed Morphology: Beyond Nanoparticle-Filled Polymers," *Chem. of Matls.,* Vol. 19, 2009, 2736–2751.

28. Havriliak, S., and Negami, J., *Polymer Science: Part C,* Vol. 14, 1966, 99–117.

Index